The Hidden Injury

Praise for –

The Hidden Injury

"Were it in my power, I would assign *The Hidden Injury: It Isn't All in Your Head* compulsory reading to all medical students in the USA."

--Yehuda Ben-Yishay, Ph.D.,
Professor of Clinical Rehabilitation Medicine New York University,
Director Brain Injury Day Treatment Program

"*The Hidden Injury* is very human, very interesting, and your concept of three people in one (pre-injury, post-injury, memories)… conveys the gravity of the adjustments you have had to overcome. The concept of "loss of self" flows throughout the entire book… your book will inform people that your experience is not unique and 'recovery' is possible given the right information and approach."

--John T. Chibnall, Ph.D, Assistant Professor of Psychiatry
St. Louis University School of Medicine, Health Sciences Center

"Your approach to the topic, which is both highly personal and yet objective… rich in details and incisive observations, serves to illustrate your thesis rather than being merely an autobiographical account of a shattering experience."

--Rev. Albert Moraczewski, O.P., Ph.D.S.T. Mag
President Emeritus, The Pope John Center

"This submission is of a quality that I would expect from a published professional."

--Heartland Writer's Conference, Sikeston, Missouri: Finalist Award

The Hidden Injury

Ethel Dimont

It Wasn't All In My Head

One woman's life changing journey
after a Closed brain Injury

Ethel Dimont

Lulu Press, Inc.
Raleigh, North Carolina, United States.

Cover designed by Kathryn Robbins

Manufactured in the United States of America
10 9 8 7 6 5 4 3 2 1

ISBN 978-0-557-68402-1

This book is dedicated to my late husband, Max I. Dimont, who taught me to view events with a critical eye, without always seeing the negative.

~~~~~~~~~~~~~~

To my daughter Gail, her husband Michael, their three sons, Mark, his wife Lisa, their children Ann and Felix, Peter, his wife Tiffany, their daughter Eleanor and my third grandson Daniel.

~~~~~~~~~~~~~~

To my "Works in Progress" writer's group, who encouraged me every Wednesday to keep writing no matter what.
Thank you!

~~~~~~~~~~~~~~

To my dear friend Judy Miller, who has been part of this journey from the very beginning. I could not have done it without your encouragement, support and friendship. You are a true blessing!

~~~~~~~~~~~~~~

To Kathy Robbins, there are no words to describe how deeply grateful I am for all that you have done for me. If not for Kathy's tenaciousness, this book would not be in print, my days would not be as bright nor would I have laughed as much.

Contents

INTRODUCTION

Until the end of the twentieth century, we rarely questioned the quality of our medical care and relied solely on the advice of physicians. The twenty-first century brought with it a new approach to health. Patients are now told not to blindly accept the decisions made by their physicians about their health, to take more responsibility, and to become proactive "consumers," as if they were purchasing a product in a department store. Does this mean we depend on our physicians when we should not? I don't think so. But what does it mean? What prompted this new approach and revolutionary change? Where is this advice coming from, and why? What effect is this having on the patient–physician relationship?

My search for answers to a misdiagnosed medical problem led to this book, written for anyone who ever sees a doctor, but I hope it will be especially useful to laypeople with medical problems that are difficult to diagnose. It arose from my need to warn others who may be heading down the same tortuous path I trod. My experience altered me irreparably and became the leitmotif of my life instead of causing only a blip in it.

Physicians, for a multitude of reasons, often base their diagnosis on personal opinions, statistics, and/or myths when causes for symptoms are difficult to discern and the "search for broken body parts" fails. The symptoms are often viewed as manifested by the patient's psychological response, even when, as in my case, there had been an injury resulting in these "unreal" symptoms.

Today, forty years after my life-altering injury, very little has changed in the way patients with this type of medical situation are treated. The long, arduous path back to normalcy inevitably involves physicians, attorneys, and insurance companies. In their uncooperative hands, chronic and debilitating symptoms developed due to the long delay in receiving proper care; many years passed before I was properly diagnosed and treated. Patients are still being "dropped into the forest without a map," as Dr. David M. Eisenberg, Assistant Professor of Medicine at Harvard described it—they end up with the same problems and similar results as I had.

Long before "proactive" became part of our medical vocabulary, when my future seemed hopeless, I resorted to actively finding answers myself. Although I learned a great deal, applying it required more knowledge and capability than I had. Indeed, at the time, it was an almost impossible task.

The Hidden Injury is, among other things, an attempt to raise a clarion call to others in the same or similar medical situations, with symptoms that are difficult to diagnose, to be aware of the potential for misdiagnosis and the importance of getting proper and adequate help early enough to prevent the serious problems that will inevitably develop if serious injuries are left untreated.

The reactions of laypeople and professionals to my manuscript have been diverse and fascinating. They have found, hidden between the lines, a subtext that may be as important and possibly even more far-reaching than originally intended. Typically, physicians, often male, had explanations that were more like excuses. They had many, seemingly logical reasons for the medical role in this fiasco but little understanding for the negative effect on the injured victim. Some attorneys, again, typically men, questioned my facts but had no difficulty finding numerous reasons to explain why things happen the way they do. On the other hand, no matter what their role in society, women usually responded with a simple statement: "I know exactly what you are talking about," or, even more often, "Doctors often think it's all in your head." They had either lived through a similar personal experience or knew someone who had. The details in their stories covered a broad range of medical situations similar to mine.

The professionals who did agree with me were those who cared for the victims after they had been discharged by their physicians. They interpreted the care I received as a classic example of what happens when physical and/or emotional problems cannot be confirmed with medical proof. Patients are then seen as being "responsible for the symptoms," implying they are either neurotic, or, when a financial settlement is involved, malingerers, as if the symptoms developed because of the victim's attitude rather than the physical reason for them. This adds a "fake" psychological dimension to the problem at precisely the point at which the patient is least capable of handling it.

My story has also generated a wide variety of opposite reactions: I was either too angry or not angry enough. I denigrated doctors or I was too kind to them. All thought I should add information to stress their viewpoints. Each reader will certainly reach their own conclusions, but I tried to present what happened as it happened. I tried to be fair, to present those experiences as they occurred, and to be as objective as possible. I will admit, it wasn't always easy.

In spite of some of the excellent medical care available today, the American public has become skeptical, and the image of the doctor has been tarnished by that skepticism. How and why this could have happened, after so many years of admiration and respect for that profession, is an important question that must be answered. This skepticism has carried over to the legal system and the insurance companies. To protect their professions and the status quo, whether for public relations or financial reasons, they criticize and blame each other for the problems, often at the public's expense.

Some may be skeptical of my story questioning my ability to recall such a detailed, verbal account of my early experience. There is a simple answer: the specific details are not from memory; they were recorded as they happened, when I was still begging for help and it was not forthcoming. After each distressing appointment, I would go directly to my typewriter. It was my way of releasing the anxiety those appointments generated. I should have dated my notes, but the idea of writing a book could not have been further from my mind.

On a more positive note, I want to pay tribute to those attorneys and medical and rehabilitation professionals who are doing research and writing articles to alert those who are not aware but should be that if proper treatment is administered when symptoms first begin to develop, they can be alleviated. It is my hope this story will, in some way, help them succeed in their efforts.

If I had not finally received proper treatment, albeit more than two years too late, I would have been incapable of doing the research that led to this book. My experience and my refusal to stop searching for answers taught me much about the handicaps and difficult decisions—for good motives as well as selfish needs—of those involved. These costly problems have a ripple effect: financially and psychologically for physicians, society, victims and their families, and most importantly, physical and emotional costs to the victim.

The purpose of this book is not to find fault. It is to try to help patients and prod those in the medical, legal, and insurance professions to acknowledge the need to change that which creates and encourages these attitudes. Admittedly, these are not simple tasks, but the necessary knowledge to initiate such change is available. If I could find the answers in respected national and international medical and legal journals in libraries, others can, too. This information written by reputable members of the medical, rehabilitation, and legal professions is ignored by all too many of their colleagues and needs to be retrieved and reincorporated into the present-day health care system. It would be a constructive first step toward eliminating a major part of the exorbitant medical costs these attitudes generate. Perhaps, just as important, it would help reverse the skepticism that permeates our attitudes toward these professions. Cooperation, concern, and fairness is required to help remove this blight and the toll it takes on society.

I am most thankful for those professionals who graciously shared their knowledge and expertise with me. I have yet to find the words to convey the depth of my appreciation to the following exceptional individuals:

John Chibnall, Ph.D., Assistant Professor of Psychiatry St. Louis University School of Medicine, Health Sciences Center

Paul N. Dukro, Ph.D., Director, Chronic Headache Program St. Louis Behavioral Medicine Institute, Health Sciences Center, St. Louis University

Yehuda Ben-Yishay, Ph.D., Director Brain Injury Day Treatment Center New York University Medical Center

Cyril H. Wecht, M.D., J.D., Coroner, County of Alleghany, Pittsburgh, PA

Robert E. Hanlon, Ph.D., ABPP, Assistant Professor of Neurology, Director, Mild Brain Injury Clinic, Diplomat in Clinical Neuropsychology Washington University School of Medicine

Russell C. Packard, M.D., FACP, Guest Editor, Seminars in Neurology, March, 1994, Topic: Mild Head Injury

Kevin M. Guskiewicz, M.D., Department of OrthopedicsUniversity of North Carolina at Chapel Hill, Guest Editor Journal of Athletic Training; Special Issue: *Concussions in Athletes*

Rev. Albert Moraczewski, O.P., Ph.D., Director, Pope Pius Center, Medical Ethics Division, Catholic Diocese, St. Louis, MO

Jack H. Walters, M.D., FRCS(C), FRCOG, DABOG, Emeritus Professor, UWO, UO. Manchester, Missouri

These professionals responded to a phone call or a letter when I contacted them for various reasons, either after hearing an opinion they had expressed on a TV program, reading a news release about them, or finding information during my research in the medical libraries. None were too busy to respond.

As a nonprofessional, my appreciation for their help and the respect they showed me is deep. Their suggestions and opinions gave me the courage, confidence, and incentive not to give up. Their positive responses and suggested clarifications of the medical data made an insurmountable task possible. Though the idea and reason for writing this book developed long before I was fortunate enough to have been in contact with these people, I am not sure I could have accomplished what I always hoped I would were it not for them. I am, however, solely responsible for the content of this book.

Cynthia Shepard, editor for nationally recognized medical authors and medical books, used the word "psychofragmented," a concise term that describes the complicated reason for the inability of the brain to function normally after a closed-head injury. It is used in this book to explain why patients should not be left with the distressing thought that their attitude is responsible for their symptoms when it is not. My deep appreciation goes to Cynthia for her suggestions on how to logically present new and complicated ideas.

My daughter, Gail Goldey, a former manuscript editor for Harcourt-Brace, used her talent to help edit the five books my husband wrote and also did so with mine.

A deep-hearted thank you to my dear friend Kathryn Robbins, who helped put this book into print. Without her tireless help and support, this book would have stayed in its "raw" form. I also want to thank her for building my website and helping me write my blogs, which is something I love to write.

Permit me one more thank you: In his usual manner, though my husband is no longer with me, he has never stopped helping me. Going back to my memories and asking myself how he would have handled a situation made finding the answer less difficult. Thank you, Max.

PART I

My Story

CHAPTER 1

Before the Accident

There is no education like adversity.

--Disraeli

The one thing life did not prepare me for was how it would change after an automobile accident on February 1, 1971—my fifty-fourth birthday, the last day I would ever be my normal, creative, productive self. It would be many years before signs of normalcy would appear again. This is the tale of what I learned, and why and how I made that healing finally happen.

Up until my fifty-fourth birthday, I had been lucky. Life had been good to me. My early years had prepared me for the difficult challenges that confront all of us as time and circumstances change. In 1933, during the Great Depression, I earned $12 a week, was married at 17, had a baby girl, was divorced, and ended up caring for my wonderful daughter alone. She made life a pleasure then and still does today. Coping with these emotional and economic experiences prepared me for new challenges.

I met a man, Max Dimont, on a blind date during World War II and fell in love with him instantly. Had he asked, I would have married him on that first date. But he didn't. We met for three more dates before he sailed for the European Theatre of Operations in the Military Intelligence Division of the American Army. My delight was unbounded when I received a letter, written even before he landed in Europe, saying "If we both feel the same way when I return, I will ask you to marry me." When he returned to the States 11 months later, he did.

Max and I were two people very much in love, but four dates is not long enough to get to know much about anyone. That didn't matter. Our love, our marriage, and our life together, even with its ups and downs, could not have been better—until after the accident, when I changed.

We started with no money, no college educations, two daughters from previous marriages, a low income, and a move to a new city. All of these should have created problems, but none did. Nothing fazed us. We were happy and positive, knew what we wanted to accomplish, and set about doing it, never realizing how successful we would be.

My Finnish-born husband spoke five languages, but as he so often said, he loved English the best. Max loved books; he wrote book reviews for our local newspaper, and gave talks at Parent–Teacher Association meetings about the negative effect crime comics had on children. His reviews and lectures on Jewish history laid the groundwork for writing a 4000-year history of the Jews entitled *Jews, God and History*. This book, the first of five, was an immediate success and made him a popular internationally recognized lecturer. His income-producing job as an industrial editor and publicist allowed him to accept the myriad invitations he began to receive to lecture, first in the United States, then in Canada, South America, South Africa, Finland, Sweden, and in Israel, where his lecture at the Weizmann Institute was the only nonscientific subject ever presented to the scientists at that internationally recognized research and teaching institution. These honors changed the course of our lives, adding much joy and excitement to it.

The popularity of *Jews, God and History* added to the success of our marriage and helped us accomplish our goals: to have a happy home, a close-knit family, many friends, two good jobs, and eventually a lecturing and traveling career. If we never accomplished anything else, that would have been enough. Our success and the unforgettable memories these experiences generated became even more important to me after that fateful accident changed our lives forever. Those memories played a major role in helping me through the dark path toward which I was heading. I was not aware until then that I had been taking my good fortune for granted.

Within the next three months, the horror of that day surfaced, slowly and steadily; but I refused to see it as permanent. Bit by bit, the self I knew deserted me while I tried desperately to hold onto her. I refused to think it an impossible task, or that it could be a permanent change. Believing that time and my effort would return me to my normal self, I clung to the idea that I could and would get well, and that all it would require was persistence and a lot of effort; both had always worked in the past when I had faced a serious challenge, why should I think they wouldn't work now?

But the realization of the error in my thinking came too late to avoid what could have been avoided—that required effort on the part of others; mine alone was not enough.

With no plans to celebrate my birthday, it started like any other day: I spent the day at the office, returned home to prepare dinner, attended a French class I was taking in preparation for our first trip out of the country, to France and Israel, where my husband had been invited to lecture. Not in my wildest dreams could I have ever entertained this idea. But life often plays tricks on us, some good, some not so good. This trip was to be one of the good ones; going to the French class was not.

Like most accidents, it could have been avoided. The driver of the car in which I was a passenger made an illegal left turn; the driver of the car that hit us was speeding. I knew we were going to be hit, but I have no memory of the collision. I was told the damage to both cars was clearly visible. The insurance companies paid the high cost of repairing the cars immediately. They were not as cooperative about paying for the permanent, hidden damage to me. It was more convenient and economically productive for them to assume I was not seriously hurt, ignore what had actually happened to me, and to view me as a malingerer interested only in getting as much money as possible. It was as if they wanted to balance the high cost of paying for the cars, which they could not avoid, by not paying for my serious injury. They instructed the driver of the car I was in "not to discuss this subject with Mrs. Dimont," and this so-called friend never would answer any of my questions. It ended the friendship.

The last thing I remembered immediately prior to the accident was that we were going to be hit. The next thing I remember, I was wondering where I was and what was wrong. Why I was flat on my back, cold, naked, wrapped in a white sheet? What could I be but a corpse?

I do not know how long it was before I knew I was not. Pain, fright, strange faces and hushed voices assured me I was still alive. Between wondering what had happened, where I was, and how I had arrived there, I passed in and out of awareness and darkness. During these alternating thoughts and frightening feelings, I heard my husband trying to assure me that everything would be okay. Just seeing him was reassuring, except that his face, as white as the sheet that covered me, frightened me as I again lost consciousness.

I do not know whether I was ever fully awake that evening or when I regained consciousness—that same night, the next day, or a week later. What I do remember was a physically distressing reaction as my body responded violently to the shock of being pressed against an icy-cold wall and a gruff voice reprimanding me for not standing up straight so they could take the x-ray. I have no recollection of what happened in between being hit, being transported to the hospital, or being transferred to the Intensive Care Unit, where I remained for three days. It was as if I had had a bad dream but could not remember any of it when I awoke.

CHAPTER 2

The Predetermined Diagnosis

A default assumption that becomes a diagnosis, made automatically, is often made without consideration of elimination or inclusion of persistent symptoms.

--Author Unknown

My earliest memories of the night of the accident are hazy and definitely unpleasant. Under the circumstances, what happened may be understandable, but not excusable.

Of four passengers, I was the only one seriously injured. Both drivers were released that same night. X-rays showed four of my five injuries: a punctured lung and fractured ribs, pelvis, and coccyx. The fifth injury, a concussion—the injury that changed my life forever—was ignored for almost three years.

Because the punctured lung was life threatening, a thoracic surgeon was assigned to the case. No one was assigned for the other injuries, but in spite of them, while still in the intensive care unit, the surgeon's instructions were that I was to walk daily. My only memory of that three-day stay in the ICU was intense pain from being forced to walk with multiple fractures and a fear of falling.

I was unaware that I had been moved out of the intensive care unit until I asked for a bedpan, and no one was willing to give me one. Not until my husband arrived that evening did I finally get a bedpan, because the thoracic surgeon's notes to the nurses in the new unit stated that I was "ambulatory." My husband was outraged, and immediately contacted my personal physician, an internist, who transferred me into the care of an orthopedist. My internist visited me daily, and although he was no longer in charge after having transferred me to the orthopedist, he charged me for each visit.

My injuries must have been serious—I was hospitalized for three weeks while constantly being assured they were not. My optimistic nature made it easy to believe, so I never let myself doubt that what I was being told was true. So the greatest concern to me at the time was not my injuries, but rather the length of time those injuries would prevent me from returning to work. When I asked the doctor, his response—eight weeks—was more distressing than the accident and the injuries put together.

My husband and I accepted as truth, for too long, daily assurances that nothing was seriously wrong. Thinking otherwise would have been stressing the negative rather than being thankful for how fortunate I was to be alive. And those assurances might have been right had proper treatment been administered based on medical data, rather than on the personal medical experiences of certain doctors. It would be years before another physician recognized the importance of a symptom preceding physicians had ignored, and the error of their persistent assurances became obvious.

My optimism acted as a physical tranquilizer, although it was somewhat less effective for relieving my mental distress about the possibility of losing my job. It had taken me eight years to achieve the title and income of an account executive: Losing that would change everything, especially our financial future. The complexity of my work made me certain that the company could not afford to go eight weeks without replacing me, so it came as a tremendous relief that they did not, and I would be able to return.

While at the hospital, pills were a daily occurrence. Learning they were for pain, I did not want to take them—I had no pain. But I was told I had no choice. Pills, of course, camouflage pain; under normal circumstances, this would have been obvious to me, but at the time, it wasn't. Without being told, I was given antidepressants, which also camouflaged my emotional response—another extremely negative effect—by making everything seem less serious. Were antidepressants really necessary? If so, why was I discharged without instructions to continue taking them, or any other drugs?

One day, while still in the hospital, I felt a strange sensation in my head that went all the way down into my neck and shoulders. In his usual way of responding to symptoms, my physician assured me it was "nothing to worry about," but nevertheless suggested we call in a neurologist: "just to be on the safe side." Even then I still was not told I had had a concussion, the too often unacknowledged injury that causes invisible nerve symptoms and strange sensations in the head, just two of the potential symptoms that develop from this type of brain injury. Disregarding their importance was responsible for the continued, unchecked development of future symptoms, which were then diagnosed as my psychological reaction to the injury.

Twenty-four hours in a hospital bed makes an hour feel like a day. Three days later, when the neurologist finally arrived, I welcomed him with a smile and a "Good morning," but I made the mistake of adding "You must be a busy man."

"I have other patients besides you" was his gruff response. In hindsight, I can clearly see that overlooking that rebuff was a mistake. That response was a clue to what that neurologist did—and did not do—that day, as well as a clue to what future experiences with him would be like. His diagnosis and what it was based on, followed me like the plague until it was finally proven wrong, years too late.

The neurologist's examination consisted of no more than a routine, perfunctory reflex test and two questions: "Do you have headaches?" and "Are you going to see your opthalmologist?' My answer to the first question was "No." To the second question, I answered, "Yes, my glasses were broken, and I will need new ones." If that doctor had bothered to check my chart carefully, he would have known I could not have had headaches—the pills I was given camouflaged the pain. According to Alexander J. Nemeth (1990), with only this "neurological screening," a physician will usually be satisfied that there is no neurological involvement and therefore no need for further evaluation.

Would that I had been wise enough to know how important the neurologist's two questions were. Not until I learned too late why he had asked these questions did their importance become obvious, but by that time the injury had become chronic. The symptom he found was "superior gaze with vertical nystagmus," a rapid involuntary oscillation of the eyeballs caused by a concussion. I would love to know how he, future doctors, my lawyer, and the insurance company could have ignored that symptom.

Instinct told me that the neurologist's diagnosis of "nothing wrong" could not be correct. Unaware of the seriousness of the situation, I did not pay enough attention to my instinctive negative response to him. That evening I told my husband I believed this man thought I was neurotic. Had I followed my instinct, the compulsion or need to write this book might never have developed; even though, many of the negative effects of this so-called diagnosis are still with me.

After a three-week stay in the hospital, I was discharged with strict instructions to use a cane until told otherwise. Elated at the prospect of being home again, I began entertaining the idea that I would be able to go back to work in less than eight weeks, but this was not to be. Instead, it was to be a beginning to a more than two-year long list of disappointments.

It soon became difficult to hold onto my life-long optimistic approach to challenges. On the first day home, a sheet of pain enveloped my head and entire body, suddenly and violently. How could this happen? There was no sign of it in the hospital. Being home could not be the reason: I was surrounded by everything I loved—my husband, my own bed, my garden, my lovely home with its beautiful art and sculptures—everything that had made my life so joyful, successful, and productive.

"Bothering" doctors was not my favorite sport, but when that strange feeling in my head, lack of sleep, and increasing pain continued for a week, fright left me no choice. My symptoms were serious enough to warrant calling the doctor, but getting him wasn't easy: his assistant had erected an invisible wall between him and his patients that was as rigid as the Berlin Wall and almost as difficult to climb. Persistence was required to overcome that barrier, but overcoming the one he had built around himself was almost as difficult. Being sure he was listening was a third obstacle, one I never succeeded in overcoming.

I got no further than starting to explain about the pain in my head when I heard his unapologetic response: "Did I forget to prescribe pain pills for you when you were discharged? Where is your pharmacy? I'll have some sent over." I never did get to tell him my other symptoms.

His cavalier response is still in my subconscious: I try not to remember the unpleasant feelings that meeting generated, but each time my symptoms return, and they often do, I am forced to remember them.

I found myself slowly reaching the point where there was no room in my brain for more disappointments. The prescribed pills were for pain, but no antidepressants were prescribed. Why? Was it another sign of his forgetfulness, or could the reason be that I should never have been given antidepressants in the first place? And those pain pills were not what my body needed: they were given to me for the wrong reason, so they could not and did not help. They may even have exacerbated the problem by camouflaging the true issue, which was not pain. Ignoring the injury had caused the symptoms.

At home, I put my energy into trying to return to a normal life. I was realistic enough to know it would be a little while before I could accomplish this, and I never gave up trying. I rested when I knew I was beginning to overdo it, but the severity of the new symptoms continued to increase and were becoming a permanent part of my personality; and nothing I did or did not do brought any positive change. My husband did the shopping and prepared dinner when he came home after a busy day at his office, always assuring me not to be upset about it.

My incapacity to do even minor things, my irritability, and an inability to sleep without constantly disrupting his sleep made it difficult for me not to feel anger. I know it's a destructive emotion, one we are constantly told to avoid, but what is one to do with anger when it comes because it is warranted? This anger, though hidden, was there, though it would be awhile before I allowed myself to acknowledge it. My husband's care, understanding, and love made living through this period bearable, and I am eternally grateful for it. Being so engrossed in my incapacities, I was incapable of conveying my appreciation for his patience and all he did. When I finally was well enough to do so, he was gone.

My daughter left her three young children with her husband in New York and came to stay with us for three weeks, hoping to lighten our burden and help me get well. This was wishful thinking, because not until we knew what my problem was could anything help. Normal, healthy responses were out of my reach, and it frightened me when my illness made me feel as if my own daughter were a stranger.

My normal, positive attitude had prevailed while I was still in the hospital, when I sincerely believed I would be well in a short time. After being discharged, I searched for reasons for these new, unexplained, intangible but major physical changes. This search became an all-encompassing, unwelcome part of my life as I tried not to lose my positive approach. But stress, jumpy nerves, new symptoms, and no relief became so pervasive that my positive attitude became more and more difficult to maintain. The state of my nerves made it necessary to take two or three hot baths a night to relieve the tension so my body could sleep.

Then another change occurred, and not for the better: A strange new sensation in my head made it feel as if bugs were chasing each other across the top of my brain. This, too, made no impression on my physician, who never discussed it or made any effort to take this change, or me, seriously.

He never gave up believing there was nothing seriously wrong. Trusting him had become increasingly difficult. Adding to this distrust was his insistence that I was ready to return to my job full time. No matter how badly I wanted to return to work, and knowing how serious a mistake returning to my fast-paced job full time would be, I imagined that by describing the nature of my work at the direct mail company I could help my doctor understand why I disagreed with his suggestion. My desk was on the first floor of a six story building, one of eight desks each with four-line phones. The acoustics of the large, high-ceilinged room enhanced the ambient noise level to such a degree that it often seemed to reach the level of a plane on takeoff. Each of the seven account executives serviced ten executive salesmen, with as many as twenty or more of their own accounts. Each account had its own deadline, and each sales draft had six or more complicated parts to be typed before the documentation could be delivered to one of the four production departments. In between, phone requests had to be answered as the caller waited.

But the complex, heavy workload I would have to handle made no impression on my physician, and my protestation went unheeded. While he would not release me from using the cane, he insisted I go back to work regardless of the fact that I would have to navigate over this six-story building still using my cane. Nothing I told him changed his mind, but neither did I change mine, and I left him no choice but to accept my decision not to return to work when he said I should.

His next comment was even more distressing, because my boss was his close friend: "I'm not going to get in between you and your boss. You'll just have to make up your own mind when you are ready to go back to work." I could not help wondering what misinformation he had conveyed to my employer. I have never found the right words to describe my reaction, but I know that "shock" and "disbelief" are terms too mild to convey what I felt. How could he believe I did not want to go back to work? He could not have arrived at a more inaccurate conclusion had he tried, but my fear was that he had conveyed this attitude to my boss, who probably believed him, without ever really discussing the situation with me.

Such an approach by practitioners in various medical fields seems to suggest that it is okay to use psychology to diagnose patients. The insurance companies must love it: It helps confirm their misconceptions that injuries caused in automobile accidents are fraudulent claims because they cannot be seen; almost as often, physicians respond the same way by not searching for the cause.

I ignored the conclusion of the neurologist, well aware it would be impossible to change his belief that I was a neurotic and a malingerer. Two years later I was vindicated—but too late for this doctor to learn the error of his conclusion.

Another layer of unconscious anger was added to that ever-growing emotion, but my brain was already overloaded, unable to cope with any more distress. Still, I returned to work a month later: symptoms, problems, and all—yet another disastrous mistake. Nemeth's explanation of the possible result of this type of "silent accusation" made by those in the medical field confirmed for me the negative impact of my decision to return to work as early as I did. He writes:

Reassurances that are not accompanied by proper counseling backfire. The most common consequence is depression. High achievers are known to be exceptionally vulnerable individuals who prematurely resume work only to be devastated by their inability to think, perform, and produce at their former level of competence. Much stress, anxiety, and loss of self-esteem could be avoided by clarifying the symptoms and providing the guidance needed to gradually return the patient to his pre-morbid pace of productivity.

Common diagnostic errors often result from seeing the patient as emotional. Labeling a patient's response "Post-Traumatic Stress Disorder," even when it occurs "only in the physician's mind and [is] never verbally stated to the patient," according to Nemeth, "is an approach that connotes the condition is exclusively of psychological causation."

The lack of proper care and an incorrect diagnosis changed me and my personality, so that the person my husband had fallen in love with was gone. Not knowing what was wrong caused more problems for both of us than if we had been told the truth at the beginning. With the proper diagnosis, I would have responded to the true problem immediately. With the wrong diagnosis, my body began developing symptoms that then became chronic. This misinformation affected not only me but everyone I loved and everyone who loved me: my husband, my daughter, even my friends. Nothing I did had any positive effect on my effort to make living with me less painful for them. Yet they remained the same faithful, loving, and understanding people they had always been, and they continued to try to help me. I think the price they paid was too high, and I doubt that I can ever adequately convey to them my appreciation or how aware I was of this.

When I was officially discharged, my doctor said I no longer needed my cane. His goodbye included some strange advice: "If you still have headaches, why don't you see the neurologist?" Having ignored my description of the strange feeling in my head, he now referred to it as a "headache," a term I had never used. Having paid no attention to this or any of my symptoms until now, what could have prompted him to suggest I go back to the neurologist now? I was so tired of not getting answers to my questions that I resisted the desire to ask him. Asking would have required a believable answer, something I did not think would be forthcoming.

Under the care of this doctor, I struggled to retain some measure of my former self-confidence. But since the accident, the seeds of doubt this doctor had implanted in me—by questioning my motives while ignoring my questions—began spreading, limiting my ability to successfully face the many new challenges I confronted. Finally, when it could no longer absorb what was happening to it, my brain rebelled. My mind and body had become incapable of doing anything well or even functioning normally, as if they were on the verge of giving up but had not quite yet done so. Intellectually, I fought desperately but unsuccessfully to hold onto the "me" of the past, firmly convinced the answer to the problem lay in my being able to find an open-minded physician who would listen to the details I used to describe "me" before the accident and "me" after and evaluate what had happened.

Not wanting to make obvious how I really felt, I camouflaged my feelings. In an effort to cover such strong emotions and the changes that were taking place, I dressed as I always had, kept my hair well groomed, and tried my best to look outwardly like my former, normal self. This strategy was successful for awhile, until others, who until now had been oblivious to the depth of the change that had taken place, began to wonder what had happened to me. My personality had changed, and this was not as easy to hide. I was no longer the friendly, attentive, and responsive person who had been so well-liked. When my face began to show the anxiety I had been trying to hide, I could no longer keep up the act. Others could now see it. When I realized this, I began to spend more time alone at home, and I became depressed about what was happening.

My normal emotional responses had slowly become abnormal, and my sexual responses seemed to be doing the same thing. The stimulation I had always responded to with pleasure now brought tears instead of an orgasm. Without the proper diagnosis, I was unable to learn what was causing these problems, making overcoming them impossible. I found this situation difficult to accept, and I didn't know what to do about it.

Though my job was waiting for me, the stress I would encounter there was even more overwhelming than I had anticipated, adding to my already crumbling self-esteem. It was not long before it was obvious to all that coping mentally, physically, and emotionally with the pressures and deadlines was no longer within my capability. The executive-functioning and decision-making qualities that had been crucial to my ability to do my work had diminished to such a degree that the cost of each transaction I handled skyrocketed. Though the changes in me were obvious to my employers and co-workers, I refused to face reality until it was no longer possible not to.

These changes were now a permanent part of my personality. Having the "new me" in control was exactly what I had been trying to prevent, but my awareness of this change was impossible to ignore. It alerted me to the fact that I was losing any connection I might still have with the "old me," and, with that knowledge, any control I thought I still had seemed to be slowly disappearing.

An important series of thoughts began to intrude on my wounded but still-thinking brain:

Was I really responsible for my symptoms?

Why did the neurologist ask me about headaches?

Why did he ask me about seeing my ophthalmologist?

Why did my doctor suggest I see the neurologist after he discharged me?

Why did he insist I return to work when he did?

Why had he never told me anything that would help me improve?

Finally, and of utmost importance, what made that doctor view me as a neurotic and a malingerer?

Doubt continued to press with chilling regularity into that part of my brain that could still think, albeit not clearly. These disturbing questions, hidden in my mind for much too long, now began to slowly manifest themselves, forcing me to face them. It soon became evident that my only option was to find the answers for myself. Totally unaware of the almost impossible task I was undertaking, this idea never left me, although it would take some time before I gave it serious consideration. It was then that I began my own serious research, using a technique I had to develop for myself, since I had never done any research of this kind before.

My escalating doubts were as destructive as a developing physical disease, limiting my ability to recognize truth from illusion. One of my editors used the term *psycho-fragmentation*, the only logical explanation I have ever found for why I went to see the same neurologist I so distrusted—another of the many serious errors I would continue to make during those early years, before I finally learned why I was making them.

UNSPOKEN ACCUSATIONS

Just because you cannot get a diagnosis, or you are faced with a physician who doesn't appear to understand the severity of what you are experiencing each day, does not mean it is "in your head." -- Sue Falkner-Wood

I could not have made a more serious error than to follow the instructions of my orthopedist to see the neurologist again. It was a most grievous mistake, another one added to a long list of inaccurate medical advice I followed. It should have occurred to me this might possibly mean going through the same experience I had gone through with him previously. Today I understand why it didn't; at the time, my brain was incapable of such understanding.

Einstein once said, the definition of insanity is doing the same thing but expecting a different result. What was it I expected from my visit to the same neurologist? No more than what all patients expect when they see a physician: a thorough examination, an explanation or understanding of the reason for their symptoms, and advice about what to do to improve. Not even my most creative imagination was capable of imagining what actually happened next.

His welcoming remark as I entered his office, "I don't remember you," should have alerted me that something was seriously wrong. Why would a physician say that? The answer came years later: A medical requirement for a neurologist when treating a patient is to take a case history; this neurologist had not taken one in the hospital, and he did not take one now. Without this, his inaccurate conclusions, both then and now, could only have been based on the diagnosis he did not record in my hospital record. According to Jon Stone, Consultant Neurologist, Edinburgh, "One of the big problems patients with functional and dissociative neurological symptoms experience is a feeling that they are not being believed. This is partly because many doctors are not trained well in physical symptoms that are not due to disease and research in these areas is very poor.

"Some doctors really don't believe patients with these symptoms. Others do believe there is a problem and are as keen to help as they would be if you had multiple sclerosis. So if it's a real condition but it's not a disease, what is it? Are you just imagining it? The answer is you are not imagining or making up your symptoms and you are not 'going crazy'. You have functional/dissociative symptoms Getting your head around this can take time. You don't have a disease, but you're not imagining it either."

In my case, unfortunately, I placed more significance on my symptoms than did this doctor. His indifference, both at the hospital and at this later evaluation, were major causes of why and how all my symptoms became chronic and my years afterward a nightmare.

The only question the neurologist asked me during that visit was, "Why did you want to see me?" My response, of course, was to go over the same list of symptoms I had given my orthopedist and the problems those symptoms were causing me. I did not see the neurologist record this information while I was talking. I expected some comment, but none followed. No questions, no comments. Silence filled the room.

In his examining room, I was faced with the last thing I could have expected: I received no more than the same perfunctory reflex examination he had administered at the hospital before returning to his office. I endured another lengthy silence while I waited patiently for him to speak. Instead, he sat silently, fidgeting through his papers. It was I, not he, who broke the silence.

"What's wrong with me?" But what I should have asked was, "Why only another reflex examination? And what could that have to do with the symptoms I just described?" Years later I finally did get answers to my questions, but not from him, and too late to avoid what might have been avoided—the symptoms becoming chronic.

When the neurologist finally did respond, he said in a cold and indifferent tone, "Forget it, it will go away." I thought I was listening to someone deliver a Western Union telegram over the phone, where the message is restricted to ten words or less to keep the cost down. I had no choice but to interpret this as his best diagnosis and proposed treatment: Forget it.

I secretly questioned my role in this situation. Was there something wrong with me? Was I supposed to understand this as a diagnosis? If so, do I just wait and do nothing to "get better?" My proactive personality would not let me do that.

My many questions met with more silence, and each speechless period seemed longer than the previous one. (If silence is indeed golden, and his fee had been based on that, I'm sure I would not have been able to afford him.) But what was I to do now? The answer eluded me. I tried to control my distress, which was slowly reaching the point where I was afraid of how my voice would sound when I finally did speak. But trying to interpret what that six-word sentence meant was not easy. I understood each word, but the meaning of the *sentence* was unclear. A sense of something horrifying inundated me as the subtle realization that I probably was not going to get any help became obvious. It was as if we were playing a game of Charades, and it was up to me to interpret what it all meant.

Not one to give up, I had come for help; and my stubborn tenacity would not easily allow me to leave without getting what I had come for. I had no choice but to forget about being polite and somehow found the courage to ask the question that needed to be asked, one I made the mistake of not asking earlier.

"Are you telling me I am neurotic?"

I had been aware of his attitude at the hospital. Although my previous questions had been so easy for him to ignore, this one seemed to have the reverse effect. His response was instantaneous. An oh-so-slight chameleon-like change came over him, as if he were unable to hide his satisfaction, and out came another cruelly calculated uninformative response: "I'm not exactly saying that."

No longer concerned about how I might sound, without trying to hide the disrespect in my true feelings, I blurted out, "Then what exactly are you saying?"

Amazingly, I got the same six-word treatment plan and prognosis: "Forget it, it will go away."

I left that office with no choice but to assume that what I thought he meant was correct. He did think I was neurotic, a malingerer. It's what his diagnosis implied. And this doctor's conclusion influenced future physicians and prevented me from getting a correct diagnosis and proper treatment until years later, when the fallacy of this conclusion was obvious to another physician, who, by listening to my story, came to understand and believe me—and wanted to help.

According to John Chibnall, PhD, "Because of the respected position held by the physician in the eyes of the general population, a comment which is made off-handedly by him and then quickly forgotten may, nonetheless, cause the patient considerable distress. Equally important, the physician must develop an enhanced awareness of potential biases toward certain subgroups of patients (e.g.: women) and the capacity of these biases to influence decisions about patient care." Marshall McLuhan observed that physicians preoccupied with psychologically evaluating patients all too often do not come up with the correct diagnosis because of their misconceptions about their patients.

It appears that this is exactly what happened to me. I wanted help so badly, but it was not to be found where I was looking. Could there have been any other interpretation possible than the one I came away with: It's all in your head? According to Dr. Nelson H. Hendler (1980), such a diagnosis often means "You've exceeded my skills as a physician," a diagnosis substituting for "I can't find anything wrong." Hendler found that 40 percent of the patients he sees suffering from chronic pain have "clear-cut documented evidence of an undetected medical problem."

With no case history—no knowledge about my work, my ethics, my lifestyle, and my accomplishments in the business world—this neurologist could not see that my symptoms were important. Again, I had gone to a neurologist but apparently I had found one who practiced psychology as the basis for his diagnosis. These conclusions, based on personal opinions instead of medical data, could only lead to disastrous consequences—and they did. By accepting this so-called diagnosis from the neurologist as valid, other physicians also disregarded the need to search for some other answer, so they didn't.

But I never stopped searching. I spent two years trying to find a logical reason for the drastic change in me, positive it was not because I was neurotic. (I am well aware of my flaws, and being neurotic is not one of them.) When I finally found a physician who said, "Tell me about yourself," one who listened carefully to what I told him and ordered the tests to confirm the correctness of his diagnosis—tests that should have been ordered when the symptoms first began to develop—I finally got a correct diagnosis and the help I had been looking for since that first visit to the orthopedist a week after I was discharged from the hospital.

Ignoring the neurologist's instructions to "forget it" prevented a complete disaster. I could no longer use my energy to concentrate on recapturing the "old me," although subconsciously I never really did give up trying to find her. I just spent less time on it in order to try to live a normal life. Success was never within my grasp; it became necessary to try to adjust to the "new me," the one I did not want to be and didn't like. But one positive result came from this change in attitude: it prevented me from falling apart completely. Had I not ignored my neurologist's "diagnosis," I probably would now be the patient of a psychologist or psychiatrist who, if that diagnosis had been accepted as valid, would still be looking for the "real" reason for my symptoms.

CHAPTER 3

My Identity and Crumbling Self-Image

"The spirit of liberty is the spirit which is not too sure that it is right."

–Learned Hand

"Not seeing a problem does not mean it does not exist."

–Allan H. Frankle

BEFORE THE INJURY

The development of my pre-injury personality started during the Great Depression of the 1930s, when unemployment was so prevalent. I was one of the lucky few to be employed, and I worked at R. H. Macy & Company, a New York department store, at a weekly salary of $12. My employer demanded boundless energy, pragmatic attitudes, and the motivation to work hard and long: I had all three. Always aware of my goals, I kept them steadfastly in mind and readily absorbed whatever was necessary to accomplish anything I set out to do.

No matter what difficulties I encountered along the way, I enjoyed life and had an upbeat personality. Problems were challenges to be overcome. Responsibilities, pressures, and deadlines stimulated me—and the fear of losing my job was an ever-present danger during the Depression. After my first marriage ended in divorce, I had little money; yet the responsibility of caring for my infant daughter helped to sharpen my skills and build resilience. These factors contributed to my physical and psychological well-being, until the results of the accident damaged them irreparably.

I started the second half of my life with optimism and a feeling of accomplishment. By the age of 50, circumstances had helped me conquer most of my weaknesses and develop many additional strengths—all important parts of my personality. Now I was the wife of a successful author who had just been invited to speak at the prestigious Weizmann Institute of Science in Israel. We both had good positions in the business world that gave us the time to live the life we loved—to write, to travel, to entertain. The job I held made it possible for me to join my husband when he was invited to other countries to lecture, a luxury that would have been unavailable under other circumstances.

My job as an account executive was exciting, and I loved it. My spare time was spent working with my husband on his books, a fascinating and deeply satisfying experience. I loved my family, and they returned that love. The future held endless possibilities for success, happiness, good health, and financial security, something my family and I had been unaccustomed to during my early life in New York.

After the accident, what remained of this positive attitude was my tenacity and determination to overcome the problems it caused. The accident was not the tragedy: My family and I could have handled that gracefully and successfully. The tragedy was not learning the true nature of the problem, why and how it changed me, how it changed everyone's lives and forced me to ask, within a few short months, "Where did the person I was *go*?" The tragedy was what went wrong, and it is the subject of this book.

AFTER THE INJURY

My ability to think clearly slowly disappeared after the accident, and the struggle to get well became a hopeless effort; it overtaxed what strength I had left and eventually ended in failure. I was not accustomed to failure, and it was not my nature to settle for it, but this was the most futile effort I had ever undertaken. Handicap after handicap enveloped me, until the original "me" whom I had known so well became buried under a thick medical and physical layer of debris. My "altered self" was encircled with this ever-present feeling that a "new" me had emerged. Who was *she*? How did she get here? *She* was so different from *me* that I barely recognized myself: *she* changed from day to day, and definitely not for the better.

I seemed to be following behind this person who had taken over. I didn't want to, but I had no choice. This change was hard to manage, and I no longer smiled, not ever—as if the ability to do so had been lost to me. The person I had become was unpleasant and had no concern for others. I could no longer relate to people, which had been one of my most admired qualities. Multitasking, another skill I had once excelled at, was impossible. I could only do one thing at a time. Entertaining guests in my home became a tiring chore instead of the pleasure it had always been. Everything seemed out of my reach. When I did attend a gathering of people, I could only acknowledge one person at a time and felt lost in the crowd.

My self-confidence had been replaced with questions: Where had the "old me" gone? Would I ever be that woman again? Was returning to the life I had once lived even possible? If I had known the sad, negative answer then, I am not sure I would have been able to proceed even to where the results of the injury were taking me.

The secondary, chronic symptoms of my injury were daily reminders of the new problems that were developing. I could not banish from my mind, even temporarily, the constant awareness of what was happening to me, afraid I would slip into a mindless chasm if I did not actively combat the changes with everything I had. But I might as well have forbidden fish to swim; I was no more successful than they would have been.

I could not control the falling-apart process that eventually became obvious to everyone. My body and mind seemed to have aged 20 years, stratified at that level, and then gone on hold. I, and everyone who knew me, wondered what could have caused such a drastic change. To paraphrase Andre Gide, what is different in each of us is the unique thing that makes you *you*. Searching for that person in me became the theme of my daily life, closing out everything and everyone else. How my husband and daughter coped with this was beyond me then and still is, even today.

My feelings and my responses to feelings—anger and joy, sorrow and lightheartedness, love and sex, laughing and crying—were no longer normal. Healthy emotions we all take for granted were gone. The result was devastating. Instead of joy, I felt sadness. I had forgotten how to express positive emotions, until one day I realized I was no longer capable of doing so. Within a few short months, the look on my face was that of a depressed person who had aged overnight. Old habits, patterns of living and coping, basic characteristics, and previously successful methods of functioning were shattered. Each task of daily life had a mental or physical problem attached to it.

These misguided emotions did not stop me from questioning, probing, and searching for a way to prove *I* was not the cause of it all. It never occurred to me to see a psychiatrist. I knew I didn't need one, and no physician ever advised me to see one. What sustained me was my faith in myself, and my best description of why *I* thought I was changing was that my brain was somehow getting the wrong signals; I never realized how correct I was in my estimation, or how close I was to understanding why this was happening and what it actually meant. A first-year medical student would know what this means, yet my doctors didn't, or at least they responded as if they didn't.

The words "getting signals" are ones a person might use to describe, in simple terms, what the neurotransmitter system in the brain does: it sends messages to the body via chemicals from one cell to another across a gap, called a *synapse*. When the brain uses the wrong chemical, the message goes to the wrong area of the body. For example, when I cried for no reason, the neurotransmitter system was causing it. It is also common medical knowledge that the neurotransmitter levels—that is, the chemical levels—can change following an injury to the brain, causing exactly what I described: getting the wrong signals. Why no doctor picked up on this problem from my description of it was and still is a mystery to me.

It seemed the laws of cause and effect were turned upside down, as if things were happening to me simply because I was there. Explanations of what might be *causing* what I felt were not forthcoming, and I am not certain anyone believed what I was telling them. Even the "old me" would not have been prepared for this: the faith and confidence I had always had in myself was slipping out of my grasp; soon it was gone. But faith alone cannot cure physical injuries in the absence of proper medical care, nor can it do away with their negative results, like proper medical care can. I continued fighting, trying to convince myself that I was not neurotic and that neither I nor my attitudes and emotions were the cause of my problems. My mantra became this:

IF I had survived the poverty during the depression of the 30s, and *that* did not make me neurotic,

IF I had survived a marriage at the age of 17 with a husband who had had an extramarital affair, and *that* did not make me neurotic,

IF I survived a divorce that left me penniless with a year-old infant daughter, and *that* did not make me neurotic,

IF I survived two ectopic pregnancies and the knowledge that I would never have another child, and *that* did not make me neurotic,

IF I survived the many diverse and enormous problems, both financial and emotional, that life had cast in my path, and none of *them* made me neurotic, why would a mere accident make me so?

These past challenges seemed to have the reverse effect on me: instead of breaking me down, they tested my mettle, my tenacity in the face of adversity, and strengthened my resolve. I refused to allow myself to believe I had lost my ability to persevere. This fierce attitude bolstered my will to continue to search for answers until they were found.

This tenacity, as I viewed it—or stubbornness, as others did—now helped me develop a survival tool, although I did not recognize it as such. I resorted to the art of self-motivation, finding "reasons" to do what had to be done to continue trying to function as I had in the past, and convincing myself no other choices were left. But the tool I found had a dual effect: the positive result was that it saved me from utter despair until circumstances forced me to return to the urgent need to do what needed to be done—find answers. Using my own unique approach helped me survive this experience, but I was not as successful with this problem as I was with previous ones. When I relied on myself and my instincts, I rarely failed. This time, I was handicapped—I was relying on other people. It would be interesting to speculate what would have happened had I followed my instinct when it warned me, during that hospital visit, to beware of that neurologist.

In retrospect, I see that I was looking for the key when I couldn't even find the door. Though self-pity had never been a part of my personality, a sort of Dorian Gray disintegration had begun to set in. The human face is a ruthless biographer. Invisible cobwebs clouded my thinking and vision, and depression showed on my face, reflecting the sorrow I had been feeling at the loss of self, the ennui that developed because of my inability to function, and the impairment of my creative abilities. I was changed.

Though the orthopedist and the neurologist did not think I was too sick to be back at my office, my co-workers were more astute. Within a few days it became obvious to all that the person who had returned was not the same one they had known before. In an effort to help, they did some of my work without my employers knowing about it. My employers were understanding and tolerant until the financial cost forced them to change their tune. I was offered a less strenuous, lower-paying job—but I lost the one I had worked so many years to achieve. Though this was only supposed to be temporary, I never did get my former position back, nor did I ever earn my higher salary again; and eventually, I was asked to leave.

At home, other problems appeared. The burden on my husband prevented him from meeting the deadline for the manuscript of his next book, one we had both been working on for more than two years prior to the accident. As a result, its publication was delayed indefinitely. Also, my lucidity began to deteriorate, and I began to feel as though I had taken tranquilizers. I hid my fright as I remembered what loving and being loved felt like, but I did not *feel* those emotions; I only *remembered* them intellectually. I was unable to do anything about that. A fight was going on inside of me, but I did not know with whom, I was so detached from healthy feelings, thoughts, and needs. The war that was being waged in my confused brain made me feel as though I was waiting to see what would happen, like a spectator who is never really a part of the action. It was as if my brain was searching for something to fill a space before nothingness filled it. I lived with the constant fear of giving up; I struggled to keep that part of me alive that could still fight, that part that had become so important for my physical and mental health.

The person I had been just a few short months before the accident, happy and thankful for my sense of accomplishment and boundless optimism for the future, was a stranger to me. I was going through the motions of living, yet not living. My world had shrunk to my office, my home and my garden, where I no longer functioned well. I started making mistakes, which never happened before. I was able to type at a rate of 100+ words a minute without a mistake, but now, I was making errors which I never had before, such as, omitting, transposing, and adding letters that didn't need to be there. Home wasn't the sanctuary it used to be. It was becoming too small, too dark, and too confining. Even my garden, in which I had loved to work, mocked me. I was not strong enough to do my normal pruning and maintenance. My garden was looking like my life—out of control. It was as if an invisible wall had been built between the former me, the present me, and the outside world. Would my world ever expand again?

What had all the earmarks of becoming a permanent catastrophe was avoided by my refusal to accept medical conclusions that I found difficult to believe. I was looking for the missing link to my past self and the trivial, everyday things that had made life so pleasurable. The intense need to get answers and to never give up the search for them was getting stronger and stronger every day, even as the "reality" of what was happening had become unthinkable: a subjective reality that masquerades as truth can be as tricky as lies. It would be years before this reality would become evident as an obvious *unreality,* but at that point in time, before the truth became clear, I had no choice but to accept where I stood.

In spite of everything, enough of my old self was still there, and still strong. This part that had remained was able to convince me that being in that dark house was not as important as was struggling to leave it. I listened to that voice in me that cried out in that dark abyss; I wanted to remain positive, so I could continue trying to get well, even if it had to be only through a sheer force of will. But with no sign of improvement, it eventually became obvious, even to me, that to continue on in the direction I was going was foolhardy.

CHAPTER 4

Renewed Search

"The United States Air Force criteria for pilots eliminates anyone who has had a head injury or history of head injury.

–Dr. Paul E. Hoffman, USAF, MC

Chief Life Sciences Division, Director of Aerospace Safety

THE SEARCH

One month after returning to work, I went to see an internist I thought I could trust. Unfortunately, he was handicapped by the standard procedure of the day, which was to accept another doctor's diagnose as true and accurate. I assumed because of the reputations of the previous doctors that they could be trusted, and it apparently never occurred to this internist to question if the earlier diagnoses were correct. It would be two years before the internist, my husband, and I faced the reality that we should have questioned all of the doctors and their conclusions.

This internist, so unlike the others, was a pillar of strength when I needed him most. His concern and positive attitude gave me the support I needed while we searched for answers. He believed my symptoms were real, unlike the other doctors who told me I was making them up to get attention. He paid close attention to my symptoms: my sleepless nights, the inability to multi-task, lack of focus, emotional outbursts, and the constant pain on the side of my head where I had been hit, not a "headache" but a localized pain. It was hard to get people to understand the difference between a pain in my head and a headache. I never had a headache. His approach to my problems helped validate that I was not dishonest about my worsening condition. He also believed that, if I was ever to get well, we had to take steps to help the body heal in order for the symptoms to be relieved. It never occurred to us that we were embarking on a Mission Impossible.

The internist was thorough in everything he did, taking my first complete case history and analyzing all my responses to his questions about my symptoms. He considered my sleep problems and how overly tired I was; the relentless pain in my head; my inability to concentrate; the fear and anxiety I felt; my sensitivity to noise and light, or rather the loss of all normal sense responses—taste, smell, hearing, touch, and sight—as well as the skewed emotional responses. And just as important, he looked at all those additional, intangible symptoms that developed as each day went by. That evaluation convinced him that I had gone back to work too soon, and he strongly recommended I give up my job, so unlike the advice of all my previous doctors, who told me I was fine and all I needed was to get back to my normal life. Again, unaware of the seriousness of my problem, I refused to entertain that idea. It had taken years of dedication, hard work, and effort for me to finally earn the salary that indicated the importance of the work I did. That acknowledgment made all my efforts worthwhile. Even more important, wouldn't giving up my job *prove* I was a malingerer—the total opposite of the work ethics I was so proud to follow? With the success of my efforts, and the popularity of my husband's book, *Jews, God and History,* our future plans were for Max to retire early, to spend his time writing and lecturing, and for me to continue working until I was ready to retire.

I wish I had been a malingerer. Had that been the case, I might have followed that doctor's advice to stay home and take it easy. But because I was not, I decided not to stop working. Reluctantly, the internist went along with my decision but insisted I take some time off to stay home; and he made me promise to rest. "It is absolutely necessary for your body to begin to recuperate," he said. In addition, he suggested I wear a neck brace, the only time a neck injury ever was implied.

My attitude was psychologically sound. Everything in my background worked in my favor. With this internist's help, I should have begun to improve—but I didn't, and even he did not understand why.

My gynecologist, who had been treating me for 25 years, reacted with anger and shock at the idea anyone would consider *I* could be a malingerer, merely acting as if I were ill. I was in his office only for a routine visit, but he immediately made a call to a brain surgeon he knew to be sure he would see me immediately. Knowing me as well as he did, he knew the orthopedist and neurologist were mistaken, and he had no doubt about how serious the situation was. He was certain I was not neurotic, and thus knew that I could not be the cause of my symptoms.

I kept the appointment with the brain surgeon because I had complete faith in my gynecologist. He had saved my life when he operated on me for an ignored ectopic pregnancy. In spite of the seriousness of that situation, he was the only doctor who would come to the house at midnight, when, from the details of why we called him so late, he recognized it to be a life-or-death situation.

The brain surgeon to whom he sent me was amiable, bright-eyed, and pleasant, with an assuring smile. Though his manner was as smooth as glass, the gentleness was only a cover, a coating just thin enough to fool anyone—and it certainly did me. However, because of my gynecologist's recommendation, I was sure he would help me, and I had no misgivings about trusting him.

He appeared to be listening intently as I responded to his opening question: "What's wrong?" Starting at the beginning, I told him what had happened and asked him the same questions I had been asking since the accident. I remember thinking I ought to make a tape recording so I would not have to repeat the stressful details over and over again. It probably would have been a good idea, especially when I was given to tearful outbursts for no apparent reason. Years later I learned that crying without reason was just another "wrong signal" my brain was sending. The doctor waited patiently as I spoke, and then encouraged me to go on. "as usual." I stressed the importance of my strong need to know the reason for these symptoms: I wanted to know what was causing them, and why I was getting worse. I was desperate to be told what I could *do* about it. I wanted my life back.

At that time, I didn't realize I was committing a serious error. I made the mistake of telling him the names of the physicians involved, my experience with them, and their diagnoses, including the intended treatment plan, "Forget it, it will go away." I thought he should have been appalled, but no. He displayed no reaction, asked no questions, and gave no comment. What did he do? While it was happening, I could not believe it. His medical approach was the identical reflex examination I had just told him I had already had twice—pin pricks, crossing my legs, rubbing something on my face—more of the same.

An unhappy *déjà vu* feeling overcame me, and I feared this was all he would do. My instinct was again at work, warning me to me take note and watch out. I had a strong urge to walk out, afraid to hear his diagnosis. My *instinct,* which I again did not follow, was once again correct. Had I paid attention, I would not have followed his advice—which led to me wasting a year, waiting to improve, doing nothing.

"Mrs. Dimont, there is nothing wrong with you. Your pains are muscular. Use heat, massage your neck, take the pills your internist prescribed, and your pain will be gone in six months."

When I heard his diagnosis, I wondered what value this "second opinion" could possibly have. How could so many incapacitating symptoms be "muscular"? How could so many symptoms warrant such inaccurate advice and ineffective help from so many respected physicians?

Even in hindsight it is difficult to know whether this neurologist was more destructive than the first one, but destructive he was, yet in a different way. The damage the first neurologist managed to do with his indifference, prejudice, and psychological conclusions were no less damaging than the problems this next doctor managed, but with a more caring and gentle manner. How could he seem so kind yet be so wrong? His unequivocal misdiagnosis added yet another burden to the problem. Another incorrect "nothing wrong" diagnosis was exactly what we did not need. The answer, and what I should have been told to do, was not yet within my reach.

Brevity and blandness may be good procedures for physicians to save time and avoid responsibilities, but they are not good medical care for patients. Since mine weren't "muscular pains," they wouldn't go away by themselves. Instead, old problems remained, new symptoms continued to develop, and all were becoming more debilitating. Continuing in my usual manner created more stress than my body could handle; passively doing nothing was exactly what I should not have been doing. Later I would discover I should have been proactively resting which would have relieved the pressure on the very area of the brain that needed time to heal.

Six months later, I foolishly kept my follow-up appointment with the neurologist, even though his original diagnosis, and what to do about it, had brought no sign of possible improvement or change. In fact, by this time, an additional extremely disturbing symptom had developed that was impossible to control—my brain caused things to appear double-exposed, like a blurred photograph that results when the camera is moved while the photo is being taken. It felt as though I needed to move my head over to one side so the double-exposure would merge. It was unnerving, and it certainly wasn't "muscular."

The brain, like a car, can still function even with damaged parts, but it does so inefficiently. My damaged brain continued to function, but like a car that has been severely damaged, it needed repairs to function normally.

Today, as I look back on my choice to see the same neurologist again, even after my instinct told me something was wrong with what he was saying, it makes no sense. It didn't even then, but my choice to do just that is clear indication of the inefficiency of the decision-making part of my brain at that time. With the knowledge I accumulated in the year following his diagnosis, it is no surprise that he did not revise it. As it turns out, he was employed by an insurance company when he took me on as a patient. To rethink his diagnosis would have meant he had been wrong in the first place, which I believe he knew; but his circumstances did not permit him to do anything about it, even if he had wished to. And at that second visit, he merely did another routine reflex examination—and made one minor change in his prognosis: "Sometimes it takes a year or a year and a half for muscular pains to disappear." It was as if it were essential for him to insist the cause for the symptoms was muscular, in spite of the new symptoms. Why the change in the time frame for healing? A year and a half later, the answer became obvious.

Thomas Szasz once said that diagnostic language is designed to make an unsubstantiated conclusion sound correct, just in case it is not. I had not yet become acquainted with the word *iatrogenic*, a term the medical profession uses to explain medical problems and symptoms that develop from a wrong diagnosis, inappropriate treatment, or medical neglect. By this time, I had become a member of that unfortunate iatrogenic group of patients without being aware of it.

Doctors can and do make decisions that turn out to be unintentional errors in judgment, but honest errors are not the same as blind neglect or malignant, personal indifference. This neurologist's actions belong in one of these latter two categories. I understand that the life of a doctor is not an easy one, but my experience, and what I eventually learned about him, forces me to conclude that this was not his first misdiagnosis—another one of my many correct conclusions. (See the chapter on insurance companies to see how this has become acceptable.)

Being on the payroll of an insurance company, knowing liability insurance as well as he did, and being aware of the fact that my injury was the result of a car accident, he had the choice to refuse to take me as a patient. If he had to honestly testify in court in my favor, it would have jeopardized his relationship with the insurance company. Medical ethic rules say a physician's responsibility, morally and ethically, is to his patient. This was the dilemma he faced that prevented him from looking for the true cause for my symptoms: finding it would have been bad for him economically; overlooking it was not.

As time passed, the probability that I would get better got smaller and smaller. It was obvious to everyone I knew that I was not only losing the battle, but was also becoming less and less capable of carrying on even minor, routine duties or retaining normal, healthy relationships with friends and family. I had become a recluse—so unlike my former self. A cluster of diagnostic labels covered me: one neurologist viewed me as a neurotic; the orthopedist viewed me as a malingerer; the second neurologist, employed by the insurance company, viewed me as a statistic; and my attorney (who I will discuss later) viewed me as "just another personal injury case." I no longer cast my own shadow. I was now a "statistical shadow" of their combined conclusions.

These negative thoughts and actions converged on me and enclosed me in fear. I even began to think I must be at least partially responsible for my problems and almost convinced myself I was. No one, during all this time, had ever come up with an explanation I felt was accurate, and I had not yet been given any advice or suggestions about what to do, or what could possibly help me improve.

With all the negative thoughts and the bulk of professional opinion weighing in against me, my confused brain could not help almost believing that both my husband and my internist, despite their seriously concerned efforts, were beginning to think it *was* me who was responsible for my symptoms. One day, in desperation, I confronted my husband with that thought. We were both glad I did. Instead of anger at what must have seemed like an accusation, his usual straight-forwardness led to the only way we could ever have learned the reason for my symptoms or discovered the possibility of ever improving. He acknowledged the thought had begun to present itself.

"Yes," he said. "I am beginning to lean more toward the physicians' views than yours. After all, what else could I think when three specialists all agreed nothing is seriously wrong with you and kept insisting you would get well. Why would I not believe them?" He continued: "I have begun to wonder what it was that made you change so drastically. You are no longer the same person I have known, lived with, and loved for almost thirty years. What if, as you have been saying, you believe their approach is based on their opinions and not on what you are telling them? It is becoming obvious we must do something about it."

When he finally had to answer his own question, "Why would I not believe them?" his anger was overwhelming. He had subconsciously believed the doctors reports instead of seeing what was happening before his eyes. His wife was not the kind of person to fake an illness of any kind. Bringing these misgivings to the surface brought back his memories of our life before the accident. We had been very close—married for almost 30 years, deeply in love, happy and successful—yet we had become, as a result of the accident, almost total strangers. I had to bring this up, even if it added to that estrangement; the circumstances left me no choice but to do so. But instead of this contention pushing us further apart, It did just the reverse—another indication of the caliber of man I married.

My husband broached the subject again the next evening. After much consideration, he made a decision, but with his usual concern for my sensitivity, he wanted me to understand how he had arrived at it. He recalled a story about a sales manager he had worked with who had had a heart attack. After months under the care of the "best" physician, but having seen no improvement, he told his doctor, "I am responsible for 175 retail stores. When a store shows numbers in the red, something has to be done. I have to tell the manager, 'Son, I love you like a brother, but your store is losing money. I know you have good reasons—a bad location, the wrong merchandise, customers are tougher here than anywhere else—but before I close the store, I've got to try a new manager.' Doc," he said, "it's the same with you. I have been under your care for months. I am not getting better. A new doctor may get me back on my feet." Within a few weeks, he was up and around, and he went on to live many more years.

After relating that story to me, Max said, "We are going to do the same thing—start all over again. My hope is that we get the same good results that man did."

My husband's decision pleased me, but still I worried. What if my doctors were *not* wrong? They were connected with the most prestigious teaching hospital in our area and one of the finest medical schools in the country. But even if they weren't wrong, I had to learn the truth.

CHAPTER 5

Finally, Some Answers

A Correct Diagnosis and What To Do Now

Blessed is the physician who takes a good case history, looks keenly at his patient, and thinks a bit.

--Walter C. Alvarez, M.D.

In Greek mythology, Sisyphus was destined to continue rolling a heavy rock up a steep hill in Hades only to have it roll down again and again. My lot was similar: I had to try to climb up a hill out of a deep depression only to keep sliding back down repeatedly. Failure is not easy to live with, but like Sisyphus, I would not give up.

My tenacity finally paid off. My husband's idea to start from scratch, and my internist's positive response to his suggestion, turned the tables. My internist was pleased with the plan and suggested a list of qualified specialists. My husband was to set up the appointments, and these new doctors were not to be given any information other than that I had been in an automobile accident. Future decisions would have to be based solely on what I told them and what their examinations showed. It was hoped that these new physicians, seeing me with new eyes, might find the answer that had not been found.

Could this decision be the beginning of my road to at least a semblance of normalcy, and the beginning of the end of my nightmare, perhaps even the path to perfect health? It did lead to answers we could not get from those physicians whose decisions had been based on subjective conclusions, instead of medical data.

My husband and my internist, fearful about what would be found and my reactions to that, were trying to prepare me for the worst. *If*— and they stressed *if*—these doctors all came to the same conclusions and agreed with the previous doctors, I would have to consult a psychiatrist or psychoanalyst. I was not sure I could handle that, but I had to face reality, no matter what that reality might turn out to be.

We went to a third neurologist, who brought the first signs of hope and lessened the despair that had become a permanent part of my life. He told my husband to be sure that I would be free for at least three or four hours to spend with him—more time than the orthopedist and the two other neurologists had spent in total. The reason for this extended visit would soon become obvious: He was to do what the others had never done: he listened long enough to take a complete case history. He would accumulate pertinent information, and he started by making decisions as if the accident had happened the day before.

This new neurologist sent me to a radiologist to see whether x-rays would show any physical damage to the brain or brain stem. However, he explained that even if "other factors" did not show up on the x-rays, it would not necessarily mean there was no possibility of a brain injury. He also suggested psychological testing to confirm whether my symptoms were caused by the brain or by other factors. I was fearful of what these "other factors" might be, but like Scarlett O'Hara, I put that off for another day.

It was my good fortune that the neurologist who had been recommended was out of town and that the neurologist who was on call for him was not taking new patients. But before my husband had a chance to call another neurologist, the neurologist who normally didn't take on new patients called—to ask him if he was the author of "Jews, God and History." My husband said that he was, and the neurologist replied, "I will see your wife."

I had never been as appreciative of the seven long years Max and I had worked on "Jews, God and History" as I was now. This neurologist's reaction to who we were and his future care made me realize how fortunate I was: it would appear that luck succeeded where diligence had paved the way. Had I given up and accepted the earlier diagnosis, I would not have had this chance, but to paraphrase an old adage, chance appears to also favor the diligent mind. Circumstances, good fortune, and this capable neurologist helped reverse my downhill slide.

When this new doctor told my husband to be sure I had at least three or four hours to spend with him, I wondered what could he possibly have had in mind. Instinct told me how different my experience with this physician would be, and as usual, instinct was right. When I met him, my first reaction was optimism: he was tall, stately, dignified, and sensitive. A light scowl was actually his concentrating mechanism, and it didn't take long to appreciate his quiet, calm, and courteous manner. His subtle humor and gentleness, characteristics that helped me relax at that first meeting, made it easier to overcome the very distressing moments I had lived through every time I repeated my disturbing story.

Unlike my previous physicians, who from the start had generated so much doubt and hostility in me with their short, hasty conclusions that seemed to have no relation to the seriousness of my problems, this doctor considered my explanations important, not the opinion of a neurotic woman. He showed respect for my integrity and accepted what I had to say at face value.

A CORRECT DIAGNOSIS

Quietly and without emotion, this neurologist opened our four-hour meeting with a simple statement: "Tell me about yourself." His reply to my question about where to begin was, "Since I know nothing about you, tell me whatever you feel like talking about, no matter when it happened or whether you think it is important or not." I realized I had his undivided attention, and began by describing what I had been like before the accident. I stressed the fear I had that I would never be myself again and how important it was to me to be able to regain my confidence and ability. I told him how the physical, intellectual, and emotional changes developed after the accident, how incapacitating they were, and the long list of questions that were never answered. No matter how hard I tried to shorten the story, I was unable to do so.

I felt incapable of altering my thoughts, and I wondered if it was because someone was finally listening. It was as if some part of me knew we were finally going to be successful in our search for the missing link and find some way to rid the body of this unending, debilitating experience. As it turned out, my ability to recap what I had been like before the injury and the way I described the development of my symptoms after that made it possible for this neurologist to diagnose the causes for them.

Because the doctor was interested in my husband's book, I talked about the joy I felt working with Max and how the book had been written. During the time we were writing *Jews, God and History*, Max would dictate to me while we drove to work, and I would transcribe my notes during my lunch hour. In the evening, Max would work from the notes while I prepared dinner. After dinner, I would type the notes with his corrections, and the process would start all over again the next day. We even limited our social life to meet deadlines, but we didn't mind, because what we were doing made up for it.

And then came the end of my "broken record" story—my thoughts, emotions, problems, and pains; the worries, fear, anxiety, helplessness and hopelessness. It all poured out in a torrent of words: how "accomplished" I had been; that now the day was not long enough, even to do less and less of what I had done before; how much anxiety this generated; the negative effect on my sexual life; and how far away I was from my former self—the me I was trying desperately to hold on to. Outwardly, everyone thought I was improving, but I knew I was not.

After almost three hours, it became even more obvious to me how much of my old self had been lost and how great the deterioration had been. The fear of hearing the truth, no matter what that might mean, and what the future might hold seemed less frightening. It was obvious I needed to learn the truth.

Though he never examined me physically, clearly this doctor had listened carefully and remembered everything I told him. This initial conversation was a thorough *case history,* the first that had ever been taken. Is it any wonder the others never found the cause of my problems?

I finally felt that someone understood what had happened to me and thought that what I had to say about it was significant. The neurologist told me it was not difficult to diagnose my problem, because I knew myself so well and was so forthcoming about describing what I had been like before the accident. I had understood and faced the reality of what I had become, and I could acknowledge it now. Changes that had been incomprehensible and frightening to me were precisely the information that contained the clues needed to make an accurate diagnosis.

But he had promised my husband and my internist he would not tell me what he found unless Max was also present, and he explained why he could not do that to me. I thought they were fearful that, if his diagnosis confirmed the correctness of the previous physicians, I would not be able to handle it. I am thankful it never came to that.

The neurologist then told me that my not knowing the truth had been more destructive than knowing it would have been. "You have been asking all the right questions and getting all the wrong answers."

He was not very happy to have to tell me what he found. His diagnosis was much more serious than anything I could have contemplated. We called my husband in from the waiting room to hear the diagnosis, and I felt grateful that we would take the news together.

And then the diagnosis: "Mrs. Dimont, you have had a brain injury." A feeling of relief overwhelmed me. I was astounded at my reaction. It did not upset me, rather I felt such relief to finally know there was a *medical reason* for my symptoms. I was vindicated; I had not been responsible for what was happening to me, and I was not neurotic. That feeling of relief has never left me.

Thomas Kay (1986) knew this when he wrote:

The most basic element in the treatment of minor head injury is identification of the problem. There is an immediate, almost magical relief, at the moment in the head injured person feels that someone has pinpointed—and really understands—the nature of [the] problem... The process of problem identification marks the starting point in the process of rehabilitation. ... Because the nature of the bruising depends on exactly where the blow occurs, the nature of the cognitive and behavioral problems will depend on what areas are damaged ... these impairments make the injury much more than "minor," nevertheless, the patient is often treated as if all major problems were resolved, and no formal strategy is implemented.

This neurologist explained the injury, how it had occurred, the nature of it, and why I was having so many problems. He had no difficulty diagnosing my problem. In fact the answer was so obvious it became even more difficult to understand, or even to excuse, why or how those other doctors could have missed it. This new neurologist was able to offer more astute medical insight, and his skill and responsibility as a physician helped change my feeling of utter despair to what felt like genuine hope.

I was relieved to hear this neurologist assure me, that without proper medical intervention I wouldn't have been able to do anything about the symptoms no matter how hard I tried. I had neither caused nor imagined the misery I had experienced. The neurologist made a special point to assure me, strongly, that I was not responsible for the misery I had caused my husband. He stressed that nothing I might have done could have prevented that. This doctor had truly listened to what hurt me, and his sensitivity to my feelings was a great comfort. The relief his words gave me was truly overwhelming.

He then explained how my symptoms fit into a pattern, like parts of a jigsaw puzzle, and they could only mean one thing: the accident had caused a serious brain injury. Had it been treated properly at the very beginning, it could have helped avert many of the secondary problems that developed. He hastened to reassure me that the accident was responsible for my symptoms, and that I was *not* a malingerer.

WHAT TO DO NOW

To convince me about what he believed had to be done now, the neurologist offered specific suggestions to help me accept that decision as correct. It was clear to me that this doctor was someone who was not only trying to help but who was well aware of the importance of my being confident that his suggestions were the right thing to do to help me improve. He warned me, "You may be letting yourself in for what might be years of treatment. After all, there is no reason to believe me any more than you did the others."

To try to alleviate any doubt I might have had about being his patient, he suggested two things: first, that I take a series of tests to see what they would show; and second, that he should not give anyone else involved in the tests any information about me, so their responses would be based solely on the results of their tests. Then, if those results confirmed his diagnosis, he would ask me to be his patient.

Neuropsychological and electroencephalographic examinations were ordered to see if the "soft" neurological signs could be substantiated by positive findings. The psychological examinations involved a number of tests: the Bender Visuomotor Gestalt Examination; the Graham–Kendall Memory for Design Test; the Minnesota Multiphasic Personality Inventory; the Reitan Series with finger tapping test, sensory perceptual examination, and trail-making test; the Rorschach technique; and the Wechsler Adult Intelligence Scale and Interview.

All tests confirmed his diagnosis—another confirmation that my extensive injuries were a perfect example of *iatrogenic damage,* the medical term used to indicate a problem that develops from an action or inaction taken by a treating physician.

With a confirmed diagnosis, at last I had a reason to be optimistic, and I could allow myself the luxury of thinking that it might be possible for my normal self to return. I felt hopeful that my heavy burden of guilt would be lightened. I had always been aware that the future does not always go the way one sees or plans it; so much more aware of that truth, I made the decision not to squander the present.

I put all my energy into getting well, and this neurologist helped by relieving my family and me of the burdens we had been living with. Proper medication, sound advice, instruction on how to cope with the secondary symptoms that had developed, and most important, a clear understanding of the reasons for my problem all helped alleviate the psychological burden that could have been avoided. With these obstacles moved out of the way, this doctor was able to gently lead me down the path that would take me as close to normalcy as I could go, to the extent my mind and body would allow it.

But overcoming the original injury, and the secondary symptoms that developed as a result of medical inaction, was no simple matter. Reversing the damage that resulted from an untreated brain injury after so long was difficult, but we had to try. Knowing the problem, understanding why it developed, and learning how to cope with it was helpful, but it could not bring about an instant cure. The results you strive for are in the distant future; too often the results you hope for are not attainable, and good results are not easy to accomplish even when you do understand the problem.

Slowly, very slowly, the nightmare I had lived with for so long did begin to fade. It never has disappeared completely, but healing did occur, and I did improve. No longer did I have to rely solely on my own resources to overcome my difficulties, I was finally being helped to help myself. Progress was slow and tedious. If only I had been told in the first place all I was learning now, some of my problems would never have developed, nor become chronic.

I viewed the path ahead positively, as a game and a challenge. Soon, I began to make progress. Getting well and clearing my mind was urgent, and slowly some of the pieces began to fall into place. My body did begin to heal, and my mind did begin to clear, although neither ever regained the place they had been. Still, I took any sign of progress as an encouraging sign.

Losing the part of me that was gone forever made the task more difficult than I could have intellectually envisioned. My life experience and the knowledge I had accumulated had helped me develop the qualities that made me an executive instead of a clerical worker, but my ability to access these qualities had been damaged. The difficulty of retrieving and reintegrating these strengths, which I reasoned must lay within arm's reach, was greater than I could have imagined. This realization led to being depressed, as it began to seem as if achieving the goal of reviving my old self was definitely not achievable. My positive optimism was put on hold.

When depression develops, it seems inconsequential because it is invisible, like some head injuries. By the time it becomes obvious to you or to others around you, the reason for the depression has been buried, making it difficult to recognize the cause. And as my emotions slowly seemed to be returning to normal, it became even more obvious how much of me had been lost since the accident. When my body appeared to show signs of being able to respond on rare occasions, it was exciting and gratifying and added optimism and hope to my view of the future. The loss of those natural feelings, and the fear that they would never return, was mainly responsible for my depression—a "disease" that cannot improve without proper medication, outside help, or both.

Under this neurologist's excellent care, some symptoms did begin to lessen, and my optimism began to return, but it was premature. Although I was able to participate in more activities, and I did recapture some of my ability to function well at my office, normal personal relationships were still not within my reach.

Thomas Kay, PhD (1986) succinctly sums up this dilemma:

We discovered, as others had reported, that these patients appeared fine until they attempted to resume their responsibilities at home, work, or school. When they did so, a significant number experienced great difficulty. They complained of inability to remember, concentrate, organize, handle a number of tasks at once, and get as much work done as efficiently as they used to. Their relationships with family, peers, and bosses often suffered, and they developed psychological problems. Their doctors were unable to find anything wrong with them, and they were thought to be having psychiatric problems or worse yet, to be malingering.

As previously mentioned, a "breakdown" during the first year had necessitated taking time off from my job. When I returned, I was assigned to another, less challenging and less responsible job at a lower salary, a change I did not find easy to accept. But I was unable to handle even this less challenging job. As Arthur Benton, M.D., Department of Neurology and Psychology, University of Iowa (1989), explains; "[W]ith patients of high educational level in an occupation which makes demands on abstract reasoning capacity, assessment of that capacity is of importance as a basis for the decision as to whether or not he is ready to return to full time work."

If I finally am able to write this book, decades after the accident that altered me forever, it is because of the excellent medical care, the advice on how to cope, and the encouragement this neurologist gave me during those years when life was so bleak. Words are inadequate to express my appreciation for his help and the deep concern he displayed at every visit.

This neurologist and my internist, are examples of what doctors are meant to be. In spite of their capabilities, only so much of the damage that resulted from an incorrect diagnosis in the beginning could be reversed. My struggle will have been worthwhile if in some small way this book helps others avoid what I and so many others have had to go through. *Closed head injuries*, as they are called, are like depression: they are difficult to diagnose, although depression is becoming less difficult to diagnose today than it was years ago. A closed head injury is especially difficult to diagnose without listening closely and paying attention to the clues only the patient can describe.

A U.S. Army World War II slogan states that, "The difficult we do immediately, the impossible takes a little longer." But difficulty is no excuse for not listening, nor is it an acceptable reason for ignoring the unusual symptoms that can develop after a closed head injury. It is also no excuse for not helping the patient help him or herself as best they can. Assuming a person is not injured but neurotic puts the problem into the irresolvable category, and "assuming" should be unacceptable by the medical, legal, and insurance professions.

The fact that doctors are also human beings is an excuse often used for an error. Human errors are acceptable, but only when the medical "legwork" has been properly done and, *in spite of that*, one comes to a wrong conclusion. A conclusion based on prejudicial assumptions that occur without compiling and evaluating hard data is not a human error— it is an outrage.

CHAPTER 6

Recovery and New Life

Although the relationship between doctor and patient
may sometimes look like a debate it is "actually a bitter
fight for survival and, like all such struggles, it is decided
not by logic but by power."

--Thomas S. Szasz, M.D.

REVERSAL OF ROLES

Patients expect the treatment they receive from a physician to be
administered in good faith and to be the best care available, even though
the diagnosis or explanation may not be what they thought it would be.
However, a patient often does not have enough knowledge to
differentiate between a "good doctor" and a "not-so-good" doctor.
Without that, it can be hard to understand the reason for the discrepancy
between the two, but one essential characteristic for a trusting and
productive relationship is good communication.

My experience with my first two neurologists is a perfect example
of this inability to differentiate. It is difficult to believe that these
neurologists did not know that a concussion does *some damage to the brain*
even if they do not know exactly what it is, and the symptoms a patient
describes may be the only clue to a serious problem. Is this why I went
through the same, but slightly different, charade with both doctors?
Because they didn't listen to the clues I gave them? Or was it because
I made the mistake of telling the second neurologist what the diagnosis
of the first had been? Did this give him an excuse to stop listening and to
ignore the concussion and its potential for late-onset symptoms to
develop? My physicians were either well aware of the concussion or
should have been, but neither placed any importance on it.

I felt a pressing need to let them know their assumptions about me
had been wrong. I was not a malingerer, and I was not neurotic. I had a

brain injury. Without telling them why, my husband and I made appointments with both doctors after I received a good diagnosis based on a thorough case history and appropriate tests. I could not chance going alone. My memory of how they had intimidated me in the past still haunted me. And I must confess: It made confronting them now a challenge.

The first neurologist escorted me into his office with the words, "I don't remember you at all." Why would he feel the need to say that, even if it were true? Did his records alert him to what he had done, or rather not done, on that first visit? Before I left his office, I had answers to questions I had lived with for years. His attitude explained most of them.

I reasoned that his notes must have included what he found when he first saw me, while I was still in the hospital: he had diagnosed me as having "superior gaze with vertical nystagmus." Had this symptom not been ignored, everyone would have known I had had a brain injury. Years later I learned about this when I saw his letter to my attorney. It contained both the diagnosis and the doctor's explanation for why he did not think that symptom was important. I was told by this doctor, but only after I asked, that his explanation was inaccurate and had no basis in medical fact. This clue was of major importance, because it would have helped avoid some of the debilitating and life-altering consequences that arose from the disregarding of its significance. This information was used to justify increased financial compensation by the legal, medical, and insurance professionals—but it could not give me back what I had truly lost.

On this final visit to that first neurologist, I began by reminding him of his diagnosis that "nothing was wrong," as if there had been no blow to my head. I reminded him that in spite of the additional symptoms I described three months later, his treatment plan was "Forget it, it will go away." I described how he had persisted in refusing to respond to my begging for any suggestion about what I could do to improve. Instead, he had continued to respond to all of my desperate questioning with that same meaningless phrase: "Forget it, it will go away."

I continued: "I knew you were wrong then, but it took two long years of deteriorating to confirm how wrong you were. Based on the same details I gave you, one of your peers knew immediately what it was—a brain injury."

His response was shocking: "It is hard to believe you have had a brain injury, you are so lucid." He flaunted his disregard of obvious medical knowledge the way a streetwalker flaunts her casualness. A first year medical student knows this is medical nonsense.

My husband, who had until now controlled his anger, finally confronted him, "Is it possible you do not know that a patient can have a stroke, which is a brain injury, and still be lucid?" I was pleased to hear Max continue, "My wife told me, after your first hospital visit, that you had ignored her symptoms, implying there was no basis for her complaints. She interpreted your response as your thinking she was neurotic, and a malingerer only interested in the insurance money that could be made because of the accident. Viewing you as a reputable physician, I made the serious mistake of attributing this to her being distressed. It never occurred to me that you could have been wrong, because you are a doctor. We paid a horrible price for that misjudgment, and here I am listening to you do the same thing. There can be no doubt about what you meant then or what you mean now. If you believe this to be the truth, why don't you state it clearly instead of implying it?" This had no effect on this man. His ability to ignore relevant medical information was again obvious.

Ignoring his *faux pas* about lucidity and a brain injury, he continued with more of the same to prove I, not he, was responsible for my problems. The brain injury was again ignored. "You must be depressed. Have you seen a psychiatrist?"

For the first time I agreed with him. "You may be right. Maybe I am depressed. Not because of what I did, but because of what you did not do. You ignored the concussion, just as you are doing now. You dismissed what you found in my eyes, an indication of a possible brain injury. You also ignored the new clues that continued to develop when I saw you three months later, never ordering any tests, and most important, responding to my pleading for advice on how to cope with symptoms that developed after my injury by repeating that same six-word prognosis: "Forget it, it will go away."

It had taken me 54 years to mature into a highly capable functioning individual. In one short, 15-minute visit at the hospital he managed to add an additional burden to those that were to develop over the next few years, adding insult to injury—quite literally—by summing *me* up as the problem.

This experience taught me that a diagnosis that implies neuroticism, even without a basis in fact, spreads in the medical community like a virus. Did my information about his error have any effect on his thinking? Not one bit! His responses indicated he continued to believe that he was right and I was wrong: he maintained that there had been nothing wrong with me, which confirmed the doubt I had always had about him.

Edward Shorter (1992) wrote: "Given the reluctance of the unconscious mind to be made a fool of, patients have always tended to reject psychological interpretations of physical symptoms… conferring upon their symptoms a kind of hopelessness."

The single-mindedness of this type of thinking leaves no room to consider true "organic" causes to symptoms, especially that of a brain injury where the injury can't be seen. If a doctor who thinks along these lines, examines a patient and doesn't find broken bones or some other recordable injury, they will default to thinking it's all in the patient's subconscious mind. Rather than speaking from his own field, neurology, he had been treating a concussion patient based on psychiatry, an area outside the realm of his specialty; an area where he had no training. As a neurologist, he should not be treating his patients as if he were a psychiatrist. This type of thinking enabled him to disregard my symptoms.

This reminds me of a joke I once heard that really isn't very funny: What do you call the person who graduates at the bottom of his class in medical school? Doctor.

Some cases of ignorance have a known cause. Preconceived beliefs are often the basis for wrong conclusions, and disregarding hard evidence is about the worst thing a doctor can do. This physician hadn't even taken a case history. Using this train of thought, as that doctor had done, and failing to gather and then *consider* the evidence that clearly pointed to the underlying *cause* for the symptoms, what other conclusion could he have come to but the wrong one? A mind that doesn't consider the facts can easily come to a false conclusion that the reason for his patient's symptoms can only be psychological in origin.

To my misfortune, I had gone to a neurologist who had stepped out of his specialty to play the role of psychiatrist. If he believed I needed that, shouldn't he have suggested that I see a qualified psychiatrist, someone who could prescribe the appropriate medication for whatever psychological "condition" now troubled me? This was one more question I did not, at the time, have the "lucidity" to ask.

At that meeting with the first neurologist, I enjoyed having our roles reversed, although I could not help feeling sorry for him. In the past, he had always spoken in a demeaning manner to me, but now he was on the receiving end of the disdain I felt for him. Subtle though it was, I could see that he put a great deal of energy into camouflaging the crack that had begun to form in his defensive armor; but he could not hide it. Listening to the anger of a victim of iatrogenic injury elaborating on his own medical carelessness left him little room to maneuver.

Years later, after I had improved enough to be able to concentrate on and comprehend the research I read, I found answers. The medical libraries I consulted contained the truth as to why and how such obvious injuries, with the type of symptoms I demonstrated, could be ignored. The answer was simple: he and the other two physicians—and even more so the third one, who examined me for the insurance company—simply ignored the data. They did not bother to read the information that I, a layperson with no medical training at all, had no difficulty finding.

Subconsciously I had hidden my anger, but I had not succeeded in dispelling it. This is likely what prompted me to respond to this unproductive meeting by sending him copies—on a regular basis, for the next 20 years—of this type of medical literature. Why? To remind him of what he had wrought and to fulfill my secret hope that my diligence might somehow prevent him from making the same mistake with someone else. Did it accomplish anything? I may never know. But years later, when he became seriously ill and was close to dying, I received a handwritten letter from him saying, "I am taking your criticism seriously." Only then did I stop sending any more information.

CONFLICT OF INTEREST

My visit with this first neurologist had been a dress rehearsal for my meeting with the second one. Unlike the negative welcoming remark of the first doctor, the second neurologist's greeting was warm and friendly. My response to his "How are you, Mrs. Dimont?" shocked him. His Cheshire Cat smile, promptly disappeared, as I responded with, "You won't be pleased. You are not going to like hearing why I am here."

Without any preamble, as clearly and carefully as possible, I began: "You knew I had been in the hospital for almost three weeks after a serious automobile accident. And in spite of all the information I gave you, you did a routine reflex examination and diagnosed the problem as muscular. The emotional and intellectual difficulties I described to you then were not important. How could a reflex examination tell you anything about the symptoms I had described? My confidence in you was generated by the fact that you were referred by my gynecologist, someone I respected. But I made the mistake of assuming your opinion would be well thought out. Was it *accidental*, on my first and second visit, that you overlooked the possibility of a more serious medical problem? Your reputation makes it difficult to assume it was *merely* an oversight."

His demeanor changed, his face reddened, as he tried to cover his agitation with a smile and, like the first neurologist, he slowly fingered his file. He had no difficulty finding what he thought was a clue, something that would take the responsibility off his shoulders and place it on mine.

He said, "My record shows you were better the last time I saw you."

"Better than what?" I asked. "Do your records also show the additional symptoms I told you had developed since that first visit? Do your records also show you did nothing more on that second visit than what you had done on the first? You paid no attention to the additional symptoms that had since developed."

Unfortunately, never acknowledging errors of past actions fosters a protective arrogance in present actions, a device that would be employed by this man to help him hide from the very real possibility of a misdiagnosis. Like all human beings, professionals also can, and often do, become careless and overconfident after years of practice. They develop a belief that their experience gives them the answers, forgetting that people and their bodies are each different and that bodies do not all respond to health problems the same way. This neurologist seemed to fit that picture.

It was no longer within his capability to respond calmly to my questions; they were too pointed. His demeanor went from shock to a blank look to an incredulous stare and then to being careless. Unaccustomed to being challenged, he continued searching his records, but this ploy did not prevent him from losing control. Without enough time to formulate answers that might have been acceptable, after a few long, silent, thinking minutes, he handed my husband his coat, opened the door, and almost pushed us both out of his office, shouting, "I don't have to listen to you! I don't have to listen to you!"

I had no intention of leaving before I could ask the one question I knew he did not want to hear or answer, the question that had been one of the primary reasons for this visit. I blocked the door and quietly asked, "Do you work for an insurance company?"

Clearly, this was not the kind of knowledge he expected his average patient would have. By the look on his face I felt he was truly surprised by this question. By accident I overheard in a conversation at a social gathering, two attorneys, discussing another insurance case, "This doctor is in the employ of an insurance company and never finds anything wrong when a patient has been in an automobile accident. He can't afford to…"

Unprepared for my question, the neurologist blurted out, as he continued trying to physically push us out of his office, "Yes." The more irritability he displayed, the calmer I became. His staff saw and heard as he lost complete control of himself.

At any other time in my life I would have considered my next behavior to be rude, but, I felt a strong need to leave this doctor's staff with the thought that he, not I, had done something reprehensible. As I walked out, I remarked, "He did not like being accused of doing something unethical, which is why he is shouting at me." I normally would have resisted saying this, but I must admit, I didn't feel bad about doing it.

Having the ethics of his actions and his responsibility to his patient versus his responsibility to the insurance company, whose payroll he was obviously on, must have pricked his conscience. I doubt whether he thought any patient ever knew this or confronted him with it.

My personal confrontations with these physicians, and the way they responded, were a reflection of the problems society, patients, and physicians were facing then and still face today. Today this "irregularity" is being discussed by physicians openly as a problem that requires correction.

The previously mentioned conflict of interest was a subject before a meeting of the Subcommittee on Health, the Subcommittee on Oversight, and the Committee on Ways and Means in the House of Representatives on March 2, 1989. During that meeting, the following was stated:

When a physician's commercial interest conflicts so greatly with the patient's interest as to be incompatible, the physician should make alternative arrangement for the care of the patient ... physician ownership in a commercial venture with the potential for abuse is not in itself unethical. (Wolinsky & Brune, 1994)

The interpretation of how to handle this question of conflict of interest is left entirely to the discretion of the physician. If we accept this solution, should this not also apply to the physician in the employ of an insurance company when an insurance case is involved, or is this not a "conflict with the patient's interest?" *

In 1998, in my search for a clearer understanding of the rules applied in this question, I wrote to the American Medical Association (AMA) and received this response from Blaire S. Osgood, staff associate of the Ethics Standard Division:

Your letter raised several concerns. The Council on Ethical and Judicial Affairs has issued policies which address these concerns. The most specific policy is entitled "Confidentiality: Insurance Company Representative. This policy reads as follows:

History, diagnosis, prognosis, and the like acquired during the physician–patient relationship may be disclosed to an insurance company representative only if the patient or a lawful representative has consented to the disclosure. A physician's responsibilities to patients are not limited to the actual practice of medicine. They also include the performance of some services ancillary to the practice of medicine. These services might include certification that the patient was under the physician's care and comment on the diagnosis and therapy in the particular case.

More general policies regarding conflicts of interest state that under no circumstances may physicians place their own financial interests above the welfare of their patients. While I realize that the scenario you described does not explicitly involve financial interests, the spirit of this policy would apply in general to personal interests of the physician. The primary objective of the medical profession is to render service to humanity; reward or financial gain is a subordinate consideration. If a conflict develops between the physician's personal gain and the physician's responsibilities to the patient, the conflict must be resolved to the patient's benefit.

It appears as if the conflict still has not been resolved in the minds of those involved, Wolinsky & Brune saying one thing and Osgood saying another. Or does my head injury prevent me from clearly understanding these answers?

I had touched on sensitive issues with these physicians by questioning their integrity, their medical ethics, and their motivation, well aware that they would be unable or unwilling to answer my questions. Their verbal responses and attitudes confirmed they were unaware of their responsibility, but that they knew what they had done.

Voltaire once said, "I disagree with what you say, but I will defend your right to say it." Many physicians who disagree with the diagnosis of another physician will remain silent about it no matter how detrimental that silence is to the patient. Aided by this "conspiracy of silence," which is still prevalent among professionals today, these individuals are allowed to employ tunnel vision to defend their conclusions.

Did I feel better after these confrontations? Not really. I would have preferred to learn that my treatment had been an isolated case, a rare, once-in-a-lifetime error. Instead, the reverse was confirmed. The accepted diagnoses, that there was no physical reason for my symptoms, resulted in physicians, attorneys, and insurance companies being able to assume my symptoms were caused by my psychological response to the situation. What other choice did they have but to view me negatively, as a neurotic and a malingerer, or as an outright fraud? Although the first two words are rarely clearly stated, but only applied, the word *fraud* is often used by insurance attorneys when explaining why they responded to claims in the way they did.

Accepting statistical hogwash as a basis for conclusions, and "using" a patient to fit them, inevitably leads to the erroneous conclusions. The results of such responses remain lodged in the minds of patients whose complaints fall on deaf ears and are never brought to the attention of the physicians responsible, which is why it was difficult for me to give up my strong need to confront the two neurologists. I did not want them to do the same to anyone else.

PART II

Barriers

CHAPTER 7

Damaging Myths

A myth is a person or thing having only an imaginary or unverifiable existence or an ill-founded belief held uncritically especially by an interested group.

—Webster's New Collegiate Dictionary

Modern myths are partial stories with just enough ambiguous information to be believable. Giving up belief in these myths requires the ability to recognize and accept verifiable data that discredit them and a willingness to acknowledge an error in judgment.

People often act upon what they believe to be true. As J. G.Frazer (1935) wrote: "Men are perpetually building theoretical castles of sand, which are perpetually being washed away by the rising tide of knowledge." The danger arises when the wisdom behind these words is ignored.

We all share in the dilemma regarding the medical profession today. It costs millions of dollars while creating unnecessary problems for those caught in its crush. The caliber and role of, and respect for, the next generation of professionals in the medical, legal, and insurance services may be at stake if the problems of today are not adequately addressed.

Among the many myths prevalent in society today, four specifically apply to health care: 1) nebulous symptoms, concussions, and closed head injuries; 2) attitudes toward gender; 3) compensation and accident neurosis; 4) malpractice.

Somewhere in the dark recesses of my mind (and it was very dark then) I had a strong feeling that the conclusions about my situation were not based on reality. My instinct told me my symptoms contained the answer to why I had changed so drastically. I knew it would be a disaster if I followed the advice to "forget it," and I refused to do so. In the end, it proved to be the wisest choice.

Further research confirmed the correctness of my conclusion and raised troublesome questions that had to be answered: What had made it possible for these highly respected professionals to come to such inaccurate conclusions? Why were these unfounded beliefs held uncritically by other interested parties? It was difficult to believe the answers to these questions could be so simple.

The rationalization for these misconceptions is not new. A comprehensive review of this situation is included in a journal article by Randolph W. Evans, M.D. (1994) entitled *The Post-concussion Syndrome: 130 Years of Controversy.* But why has this controversy continued for so long? For years, papers written by professionals in the medical, legal, and rehabilitation fields on this very subject are being published to expose the fallacy behind the controversy. Though my experience occurred 30 years ago, many professionals still believe, even if there has been a blow to the head, that there is no physical basis for the symptoms that develop after the injury. When a financial situation is involved, it is viewed as being a psychological response of the victim to the injury, and it is perceived that the symptoms will continue until the financial situation has been resolved. In other words, it is accepted that the patient is a malingerer.

Regardless of the rationalization for these misconceptions, holding on to this "ill-founded belief" limits a person's ability to avoid such error. Knowledge is accumulated through the willingness to admit that you don't know and the desire to learn the truth. In spite of the volumes written about the subject—concussion, possible brain injury, headaches, inability to sleep, loss of taste and smell —these symptoms are talked about and published in reputable medical journals, for the most part are ignored.

Myths do not develop out of thin air. They are not preplanned but originate out of need: but whose? Some myths eventually do disappear in the face of legitimate questions. Why has this one persisted? Is it a form of wishful thinking to believe that Alexander Nemeth's conclusion was correct when he wrote:

The professional/scientific model used for determining disability in cases of head trauma will inevitably change. This affects the professions that provide care to the injured individuals, as well as the courts and other legal tribunals that adjudicate their claims. Infused by new knowledge, the traditional system is slowly yielding to a more comprehensive understanding of the neurological and psychological effects of cerebral concussion. A broader conceptual framework for appraising personal injury resulting from head trauma has become necessary.

Is it his dual training in medical care and law that enables Nemeth to see that "Courts must also adapt to these developments." He quotes Dean William Prosser in *Handbook of the Law of Torts 3*, (4th ed. 1971): "The law of torts is anything but static and the limits of its development are never met." He continues: "The principles of fairness and justice will be well served if judges, as well as lawyers from either side of the adversary system, use the power inherent in their office to facilitate the change."

The Myth That "Nothing Is Wrong" after a Concussion

During those long years, as I tried to understand what was happening to me, my instinct told me the answer could only come from someone who would pay attention to my symptoms, someone who would listen and think about what I was describing, who might possibly know the reason for what was happening to me.

More than 30 possible symptoms can develop after a brain trauma or "mild" concussion. Whether any of these do develop depends on which area of the brain is injured and the lifestyle of the victim. The list that follows includes those I developed, and lived with, until I was diagnosed correctly. Is there any other injury or illness that develops as many clues to its existence in such a short period of time?

These clues break down into four areas of normal human behavior: 1) physical, 2) cognitive, 3) psychosocial, and 4) miscellaneous. The following are the most often acknowledged; they have, with minor additions, been extrapolated from E. Marcus Davis's article "Mild to Moderate Brain Injury: A Silent Epidemic," from the November 1990 issue of the legal journal *Trial*.

Physical symptoms develop early, but are viewed as temporary. They include headache, lack of coordination, altered senses of hearing and touch, and sensitivity to light and sound.

Cognitive symptoms take longer to recognize, because they only show up when the patient begins to perform normal activity, like returning to the work force. These symptoms include memory deficits, concentration problems, slowed thinking, problems with perception (sequencing and judgment), communication, and impaired reading and/or writing skills.

But I believe the *psychological* symptoms are the most devastating, because they lead to a warped view of the self. And after much time spent waiting for an appropriate diagnosis, it's no wonder that a person may be given to a little warped thinking, after so many years of being forced to live a warped life.

Because these symptoms do not develop immediately, but over a period of time, they are often viewed as subjective, causing major dysfunctions that slowly alter one's personality and capability. The persona you have known all your life is gone, and "you" are replaced by a stranger. Creating a new life, and learning to live with this bizarre stranger, is a time consuming and almost insurmountable task—not only for the injured person, but for their family, friends, and business associates. Facing this reality and accepting these drastic changes is essential in preventing it from becoming a major tragedy.

The symptoms contributing to this drastic change include behavioral dysfunction, emotional instability, fatigue, loss of empathy, depression, anxiety or nervous tension, sexual dysfunction, lack of motivation, and emotional liability or volatility, such as excessive laughing, crying, and a general difficulty relating to others.

The question again presents itself: Is the dictionary explanation for a myth, "an ill-founded belief held uncritically especially by an interested group," the answer to why so many clues could have been disregarded? Many other symptoms—among them unusual sleep patterns, loss of the senses of taste and smell, sensitivity to light and sound—are such subtle changes that it takes years and circumstances to make one aware of their importance.

My first symptom, as previously mentioned, was a serious sleep problem. This was never viewed as relevant, along with the other symptoms, and so it was ignored by my physicians. Is it possible that they really did not know that "sleep–awake patterns following head injury differ from sleep–awake patterns prior to head injury ... [with] significantly decreased sleep quality" or that "sleep disorders are relatively common occurrences after brain injury... Unfortunately, there has been minimal attention paid to this common and often disabling sequelae of brain injury." -- Clinchot DM - Am J Phys Med Rehabil.

Typing errors were a change I never viewed as important, and I never did include it in my list of symptoms. This is because patients also believe in myths: If there is no pain, I am, or will soon be, okay." Only one who could think *logically* would realize, when such drastic changes occur, they must, in some way, be related to the accident. The word *logical* could hardly describe me at that time.

These typing errors were discovered months later, when my husband was proofreading copy I had typed for one of his books and noticed that the errors I made were all *transposed letters*. I was typing "hte" for *the*, "onyl" for *only*, and "evrey" for *every*. Today, twenty-five yaesr later, I still make these same errors. (No, this was not deliberate.)

Serendipity supplied the explanation for this problem. My research took me to the director of the Epilepsy Foundation in St. Louis, who explained why this was happening: "You cannot avoid this. It is your brain sending confusing signals to your fingers."

Knowing that nothing I could have done would have changed this did not make it less disturbing; but it did, and still does, relieve the stress it generated. Was my head injury also the reason why I began to confuse right and left, which I still do? This is usually not serious, just annoying: you turn the water faucet to off when you really meant to turn it up higher, or vice versa. However, this confusion is dangerous when you do the same thing while driving a car.

Stress is a symptom the medical society *stresses* is important to control. This is excellent advice one cannot quarrel with. However, when a misdiagnosis is the reason for the stress, telling patients to change their attitude and lifestyle is a waste of time. No matter how much energy patients put into this effort, it will not be productive. Expending energy to ameliorate some symptoms will be in vain. With no sign of improvement, the lack of progress will only add to the stress.

How much more productive would it be if the symptoms were acknowledged for what they were, and patients were told how to "respond" to them as they *begin* to develop? It would help control the symptoms, instead of allowing them free reign and thus become chronic. Treating the symptoms properly as they develop would result in less stress and less anger than attempts to overcome symptoms that have been allowed to develop over a number of years, especially after spending years trying to cope, often unsuccessfully, having searched for elusive answers regarding what to *do* about the symptoms. Knowing that you've spent years to find something that could have prevented the permanent damage if only you'd found it earlier creates the kind of stress that is difficult to dispel.

My symptoms were more than enough to prevent me from functioning even close to the same level I had prior to the injury. Years of stress led to depression, and these symptoms substituted for my former capabilities until I learned the reasons for them. Each symptom ignored added to the development of additional, secondary symptoms until it was impossible to separate the original symptoms from the later ones. How often I wish I had been told *what was happening to me* when the symptoms began. I would have started immediately doing whatever was possible to repair the damage, instead of having to do so years later, when my energy level was so low it had to be built up before I could even start.

The Myth about the Genders

Nowhere is the unwillingness to revise one's thinking more obvious than in our attitudes toward the differences between the sexes. Though this is an age-old dilemma, not until recently was it recognized as a myth.

It is no myth that men and women are different. Only a fool would quarrel with that. But medical care based on differences when treating the genders is something with which one could take issue, especially when the basis for disparate treatment is not based on medical data but on personal beliefs, bias, and myths.

In England, in the 1800s, though men and women suffered the same illness, they were cared for differently. Men were urged to seek health through travel and outdoor activity; women were told to find aid and solace within the family, while keeping up their wifely duties (*New York Times Book Review,* 4/3/94).

In *Office Gynecology* (1971), Dr. J. P. Greenhill observed that "many women, wittingly or unwittingly, exaggerate the severity of their complaints to gratify neurotic desires." He also recommended that doctors look for "personality factors" in the diagnosis of such conditions as menstrual disorders, menopause, urinary difficulties, sterility, and low back and pelvic pain. He also noted that the inability to menstruate "may occur in women who consciously or unconsciously cannot accept womanhood."

Although 75 to 88 percent of pregnant women experience nausea during pregnancy, the February 1973 *New England Journal of Medicine* classified it with neurosis. The 1972 textbook *Gynecology and Obstetrics, Current Diagnosis and Treatment* states that nausea "may indicate resentment, ambivalence and inadequacy in women ill-prepared for motherhood."

A 1973 survey of female students in 41 medical schools reported their lecturers frequently refer to women as "hysterical mothers," "old ladies," and "hypochondriacs," whom doctors must manage.

The explanation for this dichotomy is answered by Leslie Laurence and Beth Weinhouse (1994) who quote the views of Drs. Barbara Bernstein and Robert Kane, about how differently women are treated than men. Women do not report their illnesses the way men do. Women are open and exhibit emotional behavior; men report their complaints with stoicism generally not present in women. How the physician *interprets* this difference lends itself to diagnosing women's symptoms as "psychosomatic" and men's as "tangible." In addition, the majority of drug testing, protocol, and disease study are done primarily on men, even with diseases and ailments that primarily affect women. Benstein and Kane also found that even a comparison of unexpressive women and unexpressive men elicited different responses from the examining physician.

On a lighter note, in 1995 an excellent example of how myths color our decisions appeared in an article in our local newspaper, entitled "Stress and Distress." The discrepancy between the solutions given to men versus women in this day and age would be laughable were it not so serious. Women, the article stated, feel they have responsibilities. They were advised to change their attitudes, not to take on so many responsibilities, and to learn to say no. The article neglected to tell them to whom they might shift those responsibilities. Men, it says, feel the need to "eat nutritious foods, get plenty of exercise and workouts." The solution for relieving their stress was to "express your anger, substitute sit-down meals for sandwiches, swap a hobby for working overtime, jog instead of playing tennis." Yes, this was a serious article. I would have loved to listen in on a conversation at the dinner table when a husband and wife discussed this article, especially after the husband was advised to "express his anger" and skip his sit-down family meal.

In this book, a misdiagnosed concussion is the primary example of a hidden injury—if only it were the only one. Unfortunately, the same results apply in other similar medical circumstances, as reported in Gina Corea's book, *The Hidden Malpractice: How American Medicine Treats Women as Patients and Professionals* (1977). She writes:

Medical schools seem to fuel the belief that emotional conflict causes many female disorders. Dr. Mary Howell...points out that lecturers refer to patients exclusively as "he," except when discussing a hypothetical patient with a psychogenic disease. They then automatically shift to "she."

Food for the acceptance of these preconceptions are fed by popular literature, TV, movies, and more recently, cable—and we all pay the price: financially, healthwise, and emotionally.

Though some of these quotes date back to the early 1970s, a major residue of these attitudes still exists. Attitudes have begun to change for the better, and medical training has begun to eliminate these biases from their curricula. These changes confirm the validity of Frazer's statement that the theoretical castles of sand that man builds are perpetually being washed away by the rising tide of knowledge. Hopefully these myths, based on "ill-founded beliefs held uncritically," will eventually be washed away in the not too distant future.

The Myth of Compensation and Accident Neurosis

The myth of accident neurosis plays a major role in diagnosing a medical problem that develops after an accident or trauma when there is no visible physical injury. When the symptoms, which are also invisible, eventually do appear, the assumption is that they are transient and will go away. Unaware of the potential for this not to happen, the patient accepts the diagnosis and waits. In addition to the original assumption is the subtle implication that the problem is in the patient's head, especially if that patient is a woman.

Terms like *Post-concussion syndrome (PCS)*, or *Post-traumatic syndrome, mild head injury*, and others are used to try to clarify the situation. Does this even come close to clarifying the original cause, or does it merely acknowledge that symptoms can and do develop after a concussion that may not have been recognized earlier?

Alexander J. Nemeth (1991) explains how the controversy surrounding the term *Post-concussion syndrome* developed to cover numerous symptoms that can develop from a concussion:

Ever since neurology and psychiatry split into two separate disciplines in the latter part of the last century, the arguments between adherents to an organic theory of causation and proponents of a 'functional' or psychogenic explanation have grown more dogmatic. The narrow, either/or approach in diagnosing the disorder became perpetuated… even in the early years of research there have been those who believed that the condition is virtually always both neurological and psychological.

How and why did this psychological view assume precedence over the organic?

Randolph W. Evans, M.D. (March 1994) attributes the acceptance of the functional, or psychogenic, explanation to Miller's well-known study, reported in 1961, which gave contemporary credence to the notion of compensation neurosis:

The most consistent feature is the subject's unshakable conviction of unfitness for work, a conviction quite unrelated to overt disability, even if his symptomatology is accepted at face value.… Another cardinal feature is an absolute refusal to admit any degree of symptomatic improvement.

Was this statement based on valid medical data, or is it another example of a belief that has become embedded in a myth? Like all myths, this statement contains an element of truth, but what it does *not* contain is even more significant. Evans continues:

Other studies have demonstrated the validity of Miller's observations but, contrary to his claims…accident neurosis was identified in only 6.8% of patients…Most plaintiffs still symptomatic at the time of litigation are not cured by the litigation.

In spite of the fallacy of Miller's conclusion, as proven by the low percentage of patients who fit this category, Evans reports that the response of neurologists, when asked for their reaction to the statement about patients with post-concussion syndrome, "Once litigation is settled, symptoms quickly resolve," almost a quarter agreed or strongly agreed. There is more than enough contradictory evidence to invalidate this attitude, yet this shibboleth persists.

A large body of authoritative research has challenged such an extreme position, derived from a highly selective medical-legal perspective…Miller contended that post-concussional symptoms are rarely observed after a severe head injury…and only in compensation situation…yet many studies around the world have demonstrated post-concussional symptoms with similar rank order of frequency in head injury patients in whom compensation-seeking was infrequent.

While compensation-seeking cannot be considered a necessary or sufficient cause of post-concussion symptoms in most head injury patients… malingering is a real issue… when the presence or extent of injustice is in doubt." (Jacobson, 1995)

Shouldn't these contradictory approaches make it important enough, no matter how difficult a task, to challenge the medical profession to devise a method to distinguish valid disorders from malingering? And it is equally difficult to distinguish between the validity of this interpretation by physicians, attorneys, and insurance companies who have no more verifiable evidence for their conclusions than patients have for their symptoms. If the conclusion of "professionals" is based upon the *belief* that financial remuneration is the reason for the symptoms the victim describes, is it not equally valid for the patient to conclude that the *bottom line* is the reason for the conclusion of some "professionals," especially in the case of insurance companies.

In the dictionary, I found an alternative meaning for malingering: to *shirk doing one's duties.* Am I stretching a point when I believe some, but certainly not all, professionals fit this description?

Russell C. Packard, M.D. (March 1994) poses some questions: "I wonder at times about this controversy in regard to Mild Head Injury. What exactly is the controversy? Is it about brain injury or is it about money? How has our litigious society contributed to the controversy?

And controversy seems to fade away when the patient with MHI is a physician."

Myths, and speculation about their origins, have been with us ever since *Homo sapiens* began thinking. Myths develop for a multitude of reasons, but they all serve a common purpose: they give human beings an explanation for the observed world and for things we cannot understand. Though myths may vary in specific details from one telling to another, the basis for the myth is always preserved. To fill the need for an answer in situations where the underlying causes are not clear, the core idea is embellished to fit the circumstances, fueled by preconceived concepts and/or personal beliefs, the lack of available information, and the misunderstanding, ignorance, or outright rejection of available knowledge.

Myths that served earlier civilizations do not necessarily serve later ones, so new myths are often spawned when situations change and questions arise that cannot, as yet, be properly answered. And though a myth may die slowly, it will inevitably be replaced by another, adapted to fit the new circumstances and needs, and used as ammunition against the new challenges. New economic structures, societies and civilizations change constantly. Myths follow this road, and prevalent myths change to fit the new needs.

In medicine, a strong belief in myths, statistics, and preconceived ideas hypnotizes and blinds. It enables the believer to disregard opposing, forward-looking data and avoid evaluating alternative possibilities. Myths serve an important and often negative role in this type of thinking.

Alexander J. Nemeth (1991) clearly delineates this medical dilemma, confirming Evans' conclusion and nullifying Miller's. "Diagnostic decisions are too often influenced by false notions perpetuated by medical folklore, such as, for instance, that if there is pending compensation, symptoms unverified by standard medical procedures must come either from a preexisting neurosis or from deliberately deceptive manipulation on the part of the patient."

He continues, quoting Benson & Blumer, 1975, "Some specialists' attitudes do not reflect an enlightened view of the syndrome either." As noted by critics, "Patients whose medical problems place them in a borderland between neurology and psychiatry are particularly susceptible to incomplete management." These unfortunate patients, discharged by their physicians with inadequate advice on how to handle their incapacitating symptoms often require rehabilitative care. He notes:

It appears only some physicians have a clear concept of clinical neuropsychology, or, for that matter, of clinical psychology as a discipline, even though the latter has been around for the major part of the century.

In the face of the plight of all the untreated victims and their families, waiting for further research findings can hardly justify idly standing by and allowing antiquated and patently biased notions to prevail in deciding the fate of these people.

Do we now have another indication of a myth as an "ill-founded belief held uncritically, especially by an interested group?"

The Myth about Malpractice Suits

Choosing which myth to write about was not an easy decision. I chose this one because, like the previous myths, it applies not only to head injuries but also to medical care in general. The myth of accident neurosis still plays a major role in diagnosing patients with serious symptoms that develop after an injury, when the cause of the symptom is not obvious. Words like *neurotic* and *malingerer,* or the implication that "it's all in your head," are embedded in the minds of some of the professionals who still accept this myth.

Recently, changes in the rules regarding the use of legal redress, is being considered by the government as a "solution"—but to what problem? Is the legal system the problem, or is the problem one consequence of a more serious issue, the core problem that needs correcting to obviate the need to resort to malpractice suits? How will limiting the result of the cause, but not the cause itself, help resolve the problem? That's like treating the symptom but not the disease.

Before 1835, medical malpractice cases were virtually nonexistent. Today, they are part of our legal system, but until the middle of the 1960s, it was extremely rare for this subject to be brought to the attention of the American public. It was ignored because, like the data on "hidden injuries," reliable statistics on medical malpractice insurance before the mid 1970s was sparse.

What happened in the 1960s to cause the increase in this "growth industry"? Did patients change? Did the legal profession change? Or does the fault lie elsewhere? Were patients' claims for proper medical care and compensation valid, or as so often presented, were they fraudulent? It was hoped this malpractice "threat" would lead to a solution to patient's search for fair treatment. It seems we now need a solution to the solution.

Where is all this leading? The three professions involved—insurance companies, physicians, and attorneys—are in on this charade. All, except those who reject this concept, want to protect their turf, do not want to be viewed as responsible for any problem, and insist the "others" are at fault. Insurance companies have solutions that serve their purpose, and attorneys and physicians have solutions that serve theirs. Some of these professionals are beginning to look for answers that will serve the patient, but little that is substantive has been done to change things.

Agreement on appropriate responses to the perceived crisis is singularly lacking. One suggestion bandied about, and now being seriously considered, is to legally limit the forms and amounts of compensation. Correcting the reason for—but not the cause of—the problem leaves the patient, or I should say victim, no choice but to resort to outside help, which, in this case, is the legal profession. And such help is often unavailable, if preexisting conditions are present that would make litigation costly or difficult. If an attorney sees no hope of realizing any profit for the time expended, little incentive exists for them to take the client's case.

Resorting to malpractice suits is an act of last resort. Yes, some claims may be fraudulent. However, if patients cannot get the help they need from the medical profession, and they are offered unfair settlements by insurance companies—or no settlement at all—it leaves them no choice but to go to the legal profession for help. Correcting the cause might be a first step toward eliminating the need for lawsuits. Changing the legal procedure will not.

Malpractice cases against physicians are based on evidence that the medical care received by the patient was inappropriate. Propaganda about malpractice suits places the responsibility for them on the plaintiff and the attorney. Rarely do we hear this about the defendant. Believing who is responsible depends on one's personal approach to the problem and the myth surrounding it. But can it always be the plaintiff?

Whose purpose do these myths serve? Is this belief based on myth or reality? Are all patients caught in this dilemma, greedy, neurotic, or a malingerer? Is it myth or reality that the legal profession, in its search for income, fans disagreements and encourages these suits, regardless of their merit? Does the insurance profession have a role in perpetuating this belief? Or are low settlement offers another major reason for patients turning to the legal profession for help?

Statistics and costs for physicians' malpractice insurance and insurance company costs are readily available. Statistics for how many valid patient cases do not end up in the hands of attorneys are not. Who might benefit from protecting this turf?

Does anyone doubt that a whiplash can occur in automobile accidents? In response to my inquiry about this injury, Bryan O'Neill, president of the Insurance Institute for Highway Safety, in 1992, responded: "If you have a genuine whiplash injury or a false one, there is absolutely no way to determine who is the fraud and who is not. Even with an x-ray, you can't necessarily find something physical."

He is correct in saying that some visible physical injuries do not show up on x-rays. He is not correct in saying that a blow resulting in a sharp whipping movement of the head cannot have caused pain if an x-ray cannot "find something physical." It not only can cause pain, it can also cause an invisible injury to the brain—and it can cause permanent damage, along with pain, that also does not show up on an x-ray.

Though one cannot question the accuracy of Mr. O'Neill's statement that "you can't necessarily find something physical," the conclusions he draws from it should be questioned. He should know that a blow to the head may not cause a visible injury but can result in microscopic and/or biochemical changes that an x-ray cannot detect. Unfortunately, we have no x-ray equipment capable of registering biochemical changes.

In letters dated from 1992 to 1994, O'Neill's responses to my various questions were answered in words and conclusions like "so-called whiplash injury ... a diagnosis not used by physicians... there has been much research since our 1973 study on the causes, prevention, and treatment of this injury ... the bottom line, however, is that it is still very difficult to reliably identify these injuries ... there are many people who are faking the symptoms in order to receive financial remuneration ... the problem is that for many of these injuries, the only symptom is the person's subjective feeling of pain." (Is describing pain ever anything but subjective?) O'Neill continues, "When this is all there is, then there is no reliable or objective way to separate fake from real injuries, neither in the short or long term."

O'Neill further notes: "Someone who gets a bump on the head that does not seem serious at the time usually will not seek immediate medical treatment... if much later that person displays symptoms that are similar to those produced by a head injury, there is no scientific basis for relating the symptoms to the earlier event." I do not think O'Neill was aware of what he wrote when he referred to "a bump on the head that does not seem serious," as that is hardly a medical reason to conclude the patient is faking the symptoms. We saw recently, in the death of actress Natasha Richardson, how a "bump on the head that does not seem serious" can have very serious consequences. For her, that bump on the head lead to her death two days later.

O'Neill's statement that "there is no scientific basis for relating the symptoms to the earlier event" may sound logical, but is it? If the patient's case history points to person with a sense of responsibility, who wouldn't make things up, shouldn't this require being at least open-minded about the validity of the symptoms, especially when it is almost standard operating procedure to prescribe pain pills that camouflage, or even produce, some symptoms (such as dizziness, sleep disturbances, and so on)?

I wonder if O'Neill was aware that what he was saying sounded like he was writing my medical case history and explaining the very reason why malpractice cases are filed. When applying this logic, physicians feed the insurance companies the very information they need to continue believing what they believe and treating these situations in the manner that helps their bottom line (see Chapter 13). When the symptoms do begin to develop, and the myth that patients are only interested in money is accepted as fact, to whom can they go for help?

Is anyone listening?

The following statements made by Mr. O'Neill did not come until he realized I would not accept his answers but was searching for realistic responses and not excuses. "We are not involved in the business of insurance. Our [Insurance Institute for Highway Safety] mission is to try to find ways to prevent injuries, and we leave questions related to treatment and diagnosis to others." I wondered if he wished he had written this letter in response to my first questions.

Mr. O'Neill's responses, like all myths, had elements of truth and examples of the seeds that feed the particular myth that fits his needs. It is interesting to note that his first letters had very strong opinions on the subject but that the study to which he referred was from 1972, so he used 20-year-old data to make his point. My research contained a copy of that study along with data that *questioned* those conclusions, information that stressed the need to find ways to differentiate between "fake" and "real" injuries. Is it possible that Mr. O'Neill, like the insurance companies whose interests he represents, was unaware of the data I had no difficulty finding, because it might invalidate their conclusions? If so, are they not "faking" to protect their bottom line by not offering fair remuneration to patients? If this is called *malingering* when a patient does it, what is it called when the insurance company does it -- business as usual?

From the irritable tone of O'Neill's last letter, I dropped the subject. His answers had confirmed what had happened to me, and they were evidence of why myths develop and why they take so long to change.

There are unfair, fraudulent claims just as there are unfair, fraudulent settlements. Though the validity of some of Mr. O'Neill's statements cannot be questioned, his explanations and intent should be.

Strauss and Savitsky wrote, as early as 1934:

There can be no denying that the present mode of handling unfortunate persons in compensation bureaus multiplies the psychic stressor and strains and complicates an already almost intolerable situation. The harshness, injustice and brutal disregard of complaints shown by the physicians and representatives of the insurance companies and their ready assumption of intent to swindle do not foster wholesome patterns of reaction in injured persons. The frequent expressions of unjustifiable skepticism on the part of examiners engender resentment, discouragement and hopelessness. (Evans, 1992)

How much of this attitude has changed in the almost 60 years since this statement was made? Very few surveys have been made, and the responses in those make it difficult to assume that attitudes are any different today.

With all the progress that has been made in medical care during the latter half of the twentieth century, it is difficult to understand why a solution has not been found to distinguish between, to use Mr. O'Neill's words, "a fake or genuine" whiplash injury or actual problems versus fraudulent ones.

Unfortunately, the myth that most people fake injuries to get money is taking longer to change than society can afford. Today, it is a major contributing factor for the skepticism of patients, displacing much of the confidence and trust once put in the medical profession. Skepticism, like a virus, has spread rapidly, from the medical profession to the legal profession and especially these days to the insurance companies.

Certain questions need to be answered by those responsible—the so-called professionals. Although we can agree that human error is understandable, we must also accept that an outright mistake, which points to a lack of training or understanding or responsibility or competence, is not the same as a human error. Mistakes are the result of carelessness, indifference, and, perhaps, even the belief in a myth rather than in the conscientious application of good science. Making distinctions is essential to serve human needs; we cannot exchange the rational in favor of an irrational need for myths. However, we can try to be more discerning about our beliefs and conclusions, especially when accepting them as truth can be so destructive.

CHAPTER 8

Language as a Barrier to Proper Diagnosis

Language was given to man to conceal his thoughts.

–Stendhal

Ludwig Josef Yohan Wittgenstein (1889–1951) is reported to have said, "We are creatures of interest and purposes. By understanding the reasons for using language the way we do, we could dissolve the deepest problems of misunderstanding by asking more questions to clarify answers we do not clearly comprehend."

In *My Fair Lady*, Eliza Doolittle sings the song "Words, Words, Words." Everyone knew the "words" were part of the game they were playing. Not until the task of educating Eliza was completed did *she* realize that, though her "educators" were playing with words, it was her life that was the "plaything."

In medicine, words are not part of a game... they represent the clues and the consequences. When not clearly understood, the results can be as shocking to the patient and the physician, as was Eliza's reaction to her role in the charade to which she had acquiesced. The concept or purpose in this drama was not the problem—the problem was Eliza's inability to comprehend how its result would affect her life.

And so it is with medical terms: their use can either help to heal or to maim the patient, and their effects can be either negative or positive. The potentially frightening but unstated diagnosis of "brain injury" was not what maimed me but the clearly stated "Forget it, it will go away" that had that effect.

Jargon is defined by Merriam-Webster as follows:

1 a: confusing and unintelligible language...
2: technical terminology or characteristic idiom of a special activity or group.

If it so happens that you are not part of the "special activity or group" mentioned—that is, if you are not a doctor—you have only to refer back to Webster's first take on the word "jargon" to see the problem with medical terms.

Neither, physicians' words or patients' descriptions of their symptoms are intentionally meant to obfuscate, but they often do. Physicians believe their diagnoses and instructions are clear. Patients believe they have accurately described their symptoms. Both believe what was said and heard was understood. Why then is there so much misunderstanding?

The brain conveys information about symptoms to and from the body, and it does this *without* words. The mind develops ideas and uses words to convey them. Verbally conveying that information to the physician is the patient's responsibility. Understanding the patient's description of that information is the physician's responsibility.

When the words used by a physician to diagnose symptoms are as foreign to the patients as their written prescriptions are, how the patient interprets that diagnosis is of major importance. And who is there to help them with this? The physician—with yet more words.

Are the words physicians use their own? Do they reflect the physician's individual conclusions, based on data and thoughtful observation, or do they merely portray the assumptions of the cultural and political climate of the period? Have they been chosen carefully to be nebulous in the right way, to illuminate to some degree while also protecting the physician's legal interests?

Cooperation between members of the medical profession has complex roots, but its role as a generally accepted ideology cannot be underestimated. Patients, however, are now asking more and more relevant questions, and they are demanding more and more clarity in the answers they receive, a change that is becoming a catalyst for encouraging physicians to examine new concepts. (With the onset of the Internet, with hundreds of search engines, finding answers to these questions is getting easier.)

Though we are a verbal society, we are not sufficiently aware of how language influences our attitudes. Do words play tricks on our minds, or do our minds play tricks with words? Do we really only hear what we want to hear? How deep is our need to interpret words to fit our beliefs, personalities, and hidden agendas? How profound is our desire to satisfy some deep-seated emotion of which we may not even be aware?

We are especially vulnerable to meaning lost amid the nuances of language when we expect the person speaking to know more than we do. When something is not understood, we often fail to consider the possibility that we may be listening to nonsense, an untruth, or an unclearly stated idea. Perhaps that is why Dr. Seuss, the famous author of children's books, said: "Two sentences in a children's book is the equivalent of two chapters in an adult book."

When symptoms develop after a closed-head injury, they may be physical, psycho-social, or cognitive, depending on the area of the brain that has been injured and the degree of the injury. How, when, and if symptoms develop depends on many factors, among them is the lifestyle of the individual. Two major contributing factors that help determine how incapacitating the injury will become are the importance the physician places on the symptoms and how well the individual learns to cope.

Words are the only means patients have to convey their symptoms and problems to the physician. Listening carefully and interpreting these words correctly is the only way a physician can come close to understanding what a patient feels. Ambiguity on the part of the patient or a physician's failure to adequately question what may be unclear makes an already difficult situation even worse.

The vital link in this patient–physician relationship is often damaged when neither party is aware of or sensitive to this situation. Language is rarely static. As societies change, words take on different meaning. In today's world, these interpretations are often alluded to as merely "semantic." In medicine, this is dangerous, and it allows all of us to avoid facing the need to clarify the differences. Specific words like *cancer, appendicitis, diabetes, emphysema,* and *pneumonia* are not a problem. Their meaning is clearly understood. But the word *concussion,* and all that it implies, is not.

In addition, patients cannot and do not describe their symptoms in medical terms. How the physician transposes these words into medical vernacular is an art. If bias or preconceived beliefs are present, the response will be subjective rather than objective, confounding the diagnosis and adding to the problem.

Intuitively, I knew my symptoms contained the only clues to ever knowing the reason for them. I described them as clearly as possible, stating exactly how and what I was feeling. Did the physicians understand what I was saying? If they didn't, why hadn't they asked more questions, instead of making assumptions that led them to conclude I was neurotic? Might I have been responsible for their misunderstanding? Had I given them enough information to allow them to discern a serious problem? Years later, I learned that my words had been more than enough to convey clues that were adequate to indicate a possible brain injury.

Patients state their problems in the only way they know how. They must, however, avoid expressing personal, subjective interpretations of their pain and distress. Physicians need to be forced to ask questions to be certain that what is being described is clearly understood. Words like *headache, overly tired, nervous, can't sleep,* and *stress* are clichés, trigger words—a form of laymen's jargon that lends itself to being easily misinterpreted as subjective by physicians.

Patients' lack of medical knowledge limits their ability to evaluate the importance of subtle disorders. A head injury often limits concentration abilities and slows information processing (Sullivan, et al. 1994). A patient may also lack communication skills. Being unaware of this handicap further compromises the patient's cognitive functioning and social behavior, especially when there has been a concussion. When the physician underestimates the importance of this limitation, the result often is a subjective response rather than one that is medically based. "The appropriate care—and therefore appropriate assessment—of these patients, and the resources available to them, becomes all the more important" (Lezak, 1995).

Verbal and interpretive dilemmas exist on both sides of this dual relationship. Both participants are responsible for being consciously aware of this to avoid the inevitable misunderstandings that will develop.

THE PATIENT'S USE OF WORDS

What is the best way to describe symptoms so physicians can accurately interpret them? I thought I had been doing just that, yet their diagnoses had no relation to my symptoms. Why? Were my clues not clearly stated? Had I not given them sufficient information to consider the possibility of a serious problem? Had I said or done something that could have given them the impression that I was causing the symptoms? The words I used were vivid descriptions of medically recognizable symptoms. Properly evaluated, they were indications of a possible brain injury, especially since none of the symptoms were present before the injury.

Even words that are not always easy to understand—words like *headache, concussion, whiplash, sleeplessness*—have the same basic meaning to patients as they do to professionals. Why are they then interpreted by physicians as clues to a psychological response rather than to a physical cause? Why do physicians and insurance companies have the same sort of response to these symptoms?

What did the doctors think it meant when I told them I would wake up every five minutes throughout the night but eventually would finally fall asleep, but for no more than 90 minutes? Why did I no longer dream? Were these unimportant symptoms?

Though my research did not turn up the reason why I woke up every five minutes, I had no difficulty finding answers for my inability to dream. Dreaming sleep is accompanied by rapid eye movement (REM), and so it is said that dreaming occurs during REM sleep, a term I had been unaware of that describes a stage of sleep associated with dreaming, usually occurring three to four times a night at intervals of 90 to 120 minutes, during which time the brain waves are fast and of low voltage. The lack of this stage of sleep after a head injury is a readily recognized symptom and an important clue to the possibility of an injury to the brain. A first year medical student would recognize this lack of REM sleep as problematic, because it is an essential function of the body to rejuvenate itself—and an absolute necessity for a person's well being and good health.

Parsons and VerBeek (1982) found that sleep–awake patterns after a head injury were different from sleep–awake patterns before the injury and that sleep quality was significantly decreased. Bettar, Guilmette, and Sparadeo (1996) found that traumatic brain injury (TBI) subjects had insomnia and pain complaints much more often than non-TBI subjects. Even when TBI patients don't experience pain, even with mild injuries, they report insomnia, and even more so than with moderate to severe injuries. These findings indicate further study in the evaluation of sleep and pain problems in the milder cases of TBI patients.

The medical community seems well aware of the connection between problems sleeping and brain injury. And according to at least one study, they are aware of physicians' reaction to this information. In the words of one group of researchers: "Sleep disorders are a relatively common occurrence after brain injury and often result in a poor daytime performance and a poor individual sense of well-being. Unfortunately, there has been minimal attention paid to this common and often disabling sequela of brain injury." (Clinchot, Bogner, Mysiw, Fugate, Corrigan, 1998)

Although that studied was done more than 20 years after my injury, the problem certainly predated it and shows clearly how these injuries were viewed then, because the problem still exists today. Little attention is given to the symptoms that develop after a closed head injury.

The consequences of what might be called *catastrophic sleep loss* are well known. Allen Rechteschaffen did a famous series of experiments in which rats were "awakened to death." Rats deprived of REM sleep died in five weeks. This statement, for one who has lived with this problem for so many years—fortunately, with less disastrous results—confirmed how little my actions were responsible for my symptoms. Once I had learned about the value of quality sleep, from a medical standpoint, the subtle implication so firmly embedded in my mind, that I was somehow responsible for my problems, could now be removed based on fact rather than my belief that it was not true.

When the body and mind function normally, one does not search for information about how and why it does so. Prior to the accident, my interest in medical subjects had always been superficial. Had I known how important this information would turn out to be, my interest would have been sparked long before the injury. Understanding why symptoms can develop after a concussion would have helped me avoid a host of problems. My hope is that by sharing my story, it will help others.

The word *headache* tops the list of misunderstood words, because it is used to described any unusual feeling in the head, not just an ache. Because the word has a specific meaning, important differences and special circumstances fall by the wayside if one reads the definition of headache in *Dorland's Medical Dictionary*. Headache is described as a "pain in the head," but includes ten different categories of headaches, including a medical one, which applied to me: "post-traumatic, occurring after trauma to the head or neck…either physical or psychogenic in nature…physical causes include…stretching or tearing of ligaments or muscles in neck and injury to cervical soft tissue." If this last sentence is basic medical information, one must wonder why a brain injury was not considered as a possibility by the medical, legal, and insurance professions.

Is this not a description of whiplash, which can cause the symptoms I had described? The term "soft tissue trauma" has been substituted for "whiplash" by the medical and insurance professions, a term less readily understood by the victim. The reason for this change seems to be based on the belief that the term "whiplash" causes the victim to believe there is a problem, even if there is none; and it supplies a reason to see the victim as a malingerer.

I never did have a headache. I had, and still do have a pain in my head where it had been hit. However, that pain was not included my complaints. That blow to my head was responsible for even more involved and more serious symptoms. Using the term *headache* to describe what I felt was a mistake—and it is one that most people in this situation make and one of the chief reasons people get the wrong responses from the medical profession when such symptoms, and others, continue to develop. This is why the words we use are so important; words must be clear and specific.

In my case, it did not take long before another serious symptom developed. Slowly, a pain spread across the crown of my head that felt like "bugs building a path and then running across it." It started at the point where my head had been hit and spread across the entire top of my head to the opposite side. I tried substituting those words when I spoke to my doctor, to describe what I had been calling a headache, yet my unique description was never viewed as relevant. Vijayan and Watson (1992) discussed local headache related to soft tissue trauma or direct damage to entrapment of sensory nerves. Could it be that this was what was happening, the entrapment of sensory nerves that lead to unusual nerve sensations that spread from the point of injury? Neither of the two neurologists ever mentioned the possibility of nerve damage, in spite of that fact that I had been hit so hard that I had a bald spot in the area of the blow. Did they decide not to consider this possibility because it would have invalidated their original conclusions?

Other subtle and seemingly insignificant changes are often ignored by people who have had closed head injuries. Perhaps the injured person believes such things are fleeting. But even a temporary loss of taste and smell, sensitivity to light and sound, and personality changes are all clear indications of a possible brain injury due to a concussion. Not until such changes become permanent, and when they have become too distressing or incapacities become overwhelming, will patients realize that these changes are important enough to address.

At the risk of boring my readers, I feel compelled to mention one more symptom. I want to add it here because it turned out to be at least as important as all the others, and maybe even more relevant. Any explanation of why this clue did not lead to some consideration of the possibility of a serious head injury must be labeled for what it was: indifference, carelessness, or personal bias. My description was an obvious clue that should have enabled anyone in any of the three professions involved—medicine, law, or insurance—to know this was a *neurotransmitter* problem resulting from a head injury: a vitally important possible cause of the abnormal responses I was describing. My accurate description of "getting the wrong signals" could only have referred to one thing: the mechanism that conveys information from the brain to the body, in other words, the neurotransmitter system. By no stretch of the imagination could this description have been misunderstood to mean anything other than what it was, as my research confirmed.

I have been successful in overcoming much of the distress surrounding many of the experiences I had that might have been avoidable. However, the thought of this outright neglect, and what the correct diagnosis could have helped to ameliorate, is not easy to erase, especially when a symptom reappears. And even forty years later, they still do.

A malfunctioning neurotransmitter system is responsible for changing normal responses to abnormal ones. It changes healthy emotions, attitudes, habits, personality, and what had been productive capabilities to malfunctioning ones. This damaged system transforms a person into complete stranger, even to themselves. It did that to me.

What injury could be more severe? It was as if I had been kidnapped and replaced with a changeling. I did not recognize myself. I was no longer "me." Someone else had taken my place: I could only watch in horror, I had no idea what to do about "her" or how to live productively with "her." The utter shock of this change limited my ability to comprehend what was happening or why. Questions did not elicit answers. My confidence in my ability to cope with the situation was gone, yet I could think of no words to convey what I felt to describe this phenomena, other than "I am getting the wrong signals." My instinct, as usual, seemed to grasp how important my choice of words would be. When I finally learned what my symptoms represented, the question that had to be answered was. What did that symptom mean to those physicians? Finding answers to this question added to my compulsion to find answers to all my questions, and it made my search all the more relevant.

PHYSICIANS' USE OF WORDS: VERBAL VS. WRITTEN

Words have a life of their own. When their meaning carries the intent of the speaker, and when the listener intuitively understands that intent, a conclusion not obvious to others can be reached. Physicians seem to have an intuitive understanding of each other's words. When accurate, this is extremely productive. But, when three physicians arrive at the same incorrect conclusion, one must wonder how this could happen.

Written words have the same potential for damage as spoken words. By sheer virtue of the fact that they have been precisely recorded, they are probably even more damning at times. And keep in mind that when communication takes place in writing, it does so without the benefit of nonverbal cues that often qualify or clarify much that is left out of the actual content of what is said.

The written, legally required reports of these professionals to my attorney and the insurance companies contained clearly stated information they never told me. The first neurologist's report actually contained a reference to the clue he had ignored—one that indicated a possible brain injury. I only became aware of this important information after seeing the report.

"I found her to have normal neurologic function…except for some impairment of superior gaze with some vertical nystagmus which I believe to be due to medication she was taking at the time."

The physician's report also included the statement, "She denied headaches." Was it carelessness when he wrote this? The drugs he referred to were prescribed by my orthopedist, who insisted I had no choice but to take Valium and Dalmane, neither of which could cause vertical nystagmus—but both could camouflage pain. How could he talk about a serious symptom being attributed to the medication I was taking and then ask me if I had headaches, when the very medication to which he referred would have covered up the pain? His phrasing in the report upset me: "medication she was taking at the time." It sounded as if I were already on something, or that I had requested the medication rather than it being prescribed for me in such terms that I felt I had no choice but to take it.

Was this an unusual or routine approach in this neurologist's practice? How many other significant symptoms did this physician routinely miss?

He also asked, "Will you be seeing your ophthalmologist?" I did not realize the implication of this question at the time, but today it is obvious that he was leaving any follow up of the vertical nystagmus to someone else while he ignored it. In the hospital records he wrote, "Ophthalmologist will follow up."

The neurologist's diagnosis, within the first ten days after the accident, was clear: there was nothing wrong. His indifference made me wonder if he thought I was a neurotic, making too much fuss about something that was inconsequential. That evening, my husband and I discussed this. We made a serious error: We decided it was best to disregard this subtle accusation. And thus another error was added to the long list of those we would make until fate stepped in and changed everything.

What acceptable reason could explain why this important data was so clearly included in the medical and legal reports but never shared with me? There must be a reason for such withholding, and it had better be good.

Although it is no longer important for my well-being, it certainly is for those who may be following unwittingly along a path similar to mine.

Three physicians disregarded the possibility of a brain injury and came to the same conclusion, all implying that I was responsible for the symptoms that developed after the injury. The meaning of their so-called diagnoses came through clearly to those for whom they were intended—other so-called professionals.

I see now why subjective phrases—"I believe," "I think," and "in my opinion"—are peppered throughout these medical and legal documents. The dictionary tells us that *belief* suggests mental acceptance without directly implying certitude on the part of the believer, *think* refers to having an opinion, and *opinion* is a belief stronger than an impression but less strong than positive knowledge. The words used are designed to insulate those using them from any kind of real accountability.

Statistically, shouldn't one out of three of these qualified professionals have come up with a correct diagnosis? Or do these words mean to convey a specific conclusion for the benefit of others of their ilk: This is not a valid claim, the patient's symptoms are not "real," payments should be kept to a minimum or denied altogether. Does this underlying sentiment explain the reason for the difference between the verbal diagnoses and the written ones? Does professional courtesy impact medical diagnoses these days?

The first neurologist's erroneous conclusions, like a chain reaction, were the basis for the medical care I received for the next three years, and it explains why the words *concussion* and *head injury* were never mentioned. The insurance company's doctor (See Chapter 13) skirted this issue three years later, well aware that he was limited, as an insurance employee, in his role to diagnose the concussion as the reason for my symptoms. In an effort to show he had not ignored the concussion, he wrote "She believes she was unconscious." This sentence indicates he knew there was some head trauma, which might imply a concussion, but refused to state it clearly.

Though the words were different, the meaning and intent of the letter from the second neurologist to my attorney were the same. He wrote: "Dr. X [to the first neurologist] assured her there was nothing wrong inside her head." Why would this statement, made by an earlier physician, be included in this letter? Was it an effort to cover the misdiagnosis? I had gone to the second neurologist for *his* opinion, not to have him weigh in on a previous one.

But even more important, never having been given any sign that my head might have been part of the problem, I never could have said this.

Why the need to include this statement: it could only have been invented, but for what purpose? Does this statement leave any doubt in the reader's mind—it leaves none in mine—that he did know the reason for my symptoms; but because he also was in the employ of an insurance company, he was unable to acknowledge what he knew. And if I had been told about the possibility of a concussion, it would have alerted me to the real reason for my symptoms, which seems to be precisely what he was trying to avoid.

In spite of the long list of symptoms that were included in this letter, all relating to the head and neck area, his verbal diagnosis was that my pain was "muscular in nature." How I could have accepted this conclusion is difficult to understand, other than we are told to follow our doctor's advice and instructions. It wasn't until the obvious became the truth that I added to the long list of questions I should have been asking at the time but did not know to ask. Also included in his letter was the physician's statement: "I do not think she has any intracranial pathology." Why wasn't this ever mentioned to me? It is inconceivable that I could have ignored it.

Is there any question that what this physician offered was an opinion? If the statement had been based on medical data, wouldn't it read "There is no intracranial pathology." One can only assume he knew there was intracranial pathology but did not tell me and that he certainly did not want to acknowledge it now. Or am I being unfair?

As I write this, and search for answers to how I could have been foolish enough to have waited for things to improve, I can only attribute the waiting to sheer ignorance. I did not know the possibilities that these doctors only discussed among themselves.

The words in these official documents were not misunderstood or misinterpreted by the medical, legal, and insurance professionals. They were understood by all to mean the same thing: this patient's problem is not a result of her injury. She's responsible, not any of us. And we shouldn't have to do anything.

The discrepancy between their spoken words and their written conclusions makes one wonder if physicians have some "seventh sense" we mere mortals do not have. Would any medical board accredit physicians as qualified when diagnoses not based on medical data are made or accepted as conclusive evidence? Would these same people accept this if it happened to them? The need to correct this charade is becoming more and more necessary, especially as we look for ways to make health care more effective.

Patients go to physicians for medical help, believing their conclusions will be based on the best available medical data. They do not go for diagnoses based on personal opinions, especially when they are not told this is what they are getting. When opinion takes precedence over diagnosis, and the two become one in the physician's mind, the potential for error and misunderstanding increases, especially when all subsequent physicians tend to lean toward the previous diagnosis as a matter of professional courtesy. And in all too many cases, as in mine, this is what happens.

Just as Eliza Doolittle's teachers had no doubt about the meaning of the words they used to explain their purpose to her, so too do professionals clearly understand the words they use. Certain words convey special meanings that encourage professional conclusions based on their implications, but these are not the words the patients hear. Why? Could it be that being specific would encourage the patient to ask questions not readily answered, or that they may ask questions the physician does not want to answer? In some cases this may be acceptable, but only if no damage is done to the patient.

Though the tests untaken might still not have supplied a clear answer, expressing opinions as medical diagnoses is not an acceptable substitute. Even when there is no cure, proper medical care and advice are essential. And all the facts should be presented to the patient, in language that is not confusing, if only to help patients help themselves by asking the right questions.

The words used to express the nebulous diagnoses I lived with during those long, painful years left a deep mark on me, my marriage, my husband, and my family—a mark that can never be erased. The tragedy of this experience was not the accident, not the injury, and not the symptoms. The tragedy is that this could have been alleviated with positive and productive attitudes by the few people who had the capability to do so: the so-called professionals. *New York Times Magazine,* January 5, 1997

CHAPTER 9

Identifying the Cause of the Problem

Research studies of mild head injury (MHI), also called traumatic brain injury (TBI), lack consensus as to a unified definition of this disorder.

-- Jerrid M. Fisher Ph.D. ABPN
-- Arthur D. Williams Ph.D. ABPN

It would be misleading to reject, as valid evidence sequelae, all those abnormal features which cannot be measured. And it is important also to avoid the logical fallacy of absence of evidence asevidence of absence of mental impairment.

-- Bryan Jennett 1972.

In my attempt to find answers to what happened to me and how it could have happened, my trips to the medical libraries yielded the answers. This is what I learned.

THE IMPORTANCE OF THE BRAIN

The brain is the most complicated organ in the human body. It signals the body to tell it what is happening from joy to pain as clues to what to do.

"Pain is literally a lifesaver, alerting the brain to physical harm. Pain is the body's smoke alarm," says Robert Coghill, a neurophysiologist at the National Institutes of Health. "Because pain is so vital, the brain gives it priority over information coming from other senses."

Medicine has made great strides through surgery, medication, and psychological help in repairing readily discernible "specific" malfunctions. It has even learned to recoup a healthy organ from a deceased or healthy person, and transplant it into another human being to replace a diseased organ.

The Closed Head Injury (CHI) presents medicine with a different set of problems. The symptoms that develop after a CHI are clues to the existence of this brain injury. They are more difficult to discern because the mechanical tools and knowledge to do so are not readily available, nor are those that are readily accepted.

A brief summary of the "duties" of the brain indicates the depth of the problem of diagnosing this injury. The brain and the nervous system receive input from various parts of the body and from the outside world. Both transmit messages throughout the body, affecting the coordination, learning, memory, emotion, and thought functions.

The brain, a three-pound maze of nerves and tissue, is composed of a dense network of approximately 100 billion nerve cells (neurons) and chemicals and is the mechanism that conveys from nerve cell to nerve cell all the signals necessary to control our mental health and enable the body to function.

The nerve cell, or neuron, is the basic unit of this system and consists of a cell body, an axon, and numerous smaller branching fibers or dendrites. Neurons are connected to other neurons by synapses attached to axons and dendrites. Chemical signals from neurons are received through the synapses.

There are three kinds of neurons found in the brain -- sensory, motor, and internuncial or associative neurons. Sensory neurons enter the brain from various receptor cells. The eyes have millions of visual receptors; the ears have receptor cells that are stimulated by sound. Other receptors record pain, temperature, and touch. Motor neurons run the reverse circuit -- from brain to muscles. Although there are millions of sensory and motor neurons, the associative neurons compose the bulk of the ten billion brain cells. It is this associative area of the brain, this seat of memory and thought, with which we are most concerned, and should be of concern when there has been a concussion.

Nerve cells are different from other living cells. They cannot divide. This difference is necessary so that the brain will have fixed, stable communication pathways to and from receptors and effectors.

Neurotransmitter chemicals lock on to receptors, set in motion a cascade of chemical events in the receiving cell. This ongoing "dance of neurotransmitter and receptors" is the intricate code that brain cells use to communicate with each other. Specific chemicals serve specific purposes. For example: Serotonin is the chemical messenger that plays a critical role in sleep, mood, depression and anxiety. Norephineprine is the chemical that triggers the "fight or flight" response. However, if the neurotransmitter system malfunctions, the right transmitter might send its signal to the wrong receptor... or vice versa. Thus, when I described my "mixed up" responses as "I am getting the wrong signals," I was describing a malfunctioning neurotransmitter problem without being aware that I was.

Today, scientists know that many people suffering from brain injuries have imbalances in the way their brains use the biochemical's the neurotransmitters require to send signals from the brain and body. Too much or too little of these chemicals may result in depression; anxiety or other emotional and physical malfunctions.

The brainstem, a vital part of this procedure, connects the spinal cord to the brain. There are numerous components to this system. The brain stem controls such vital functions as heart rate, breathing and circulation of blood or blood pressure, and regulates temperature. The cranial nerves of the brain travel through the brain stem to control muscles in the face, eyes, tongue, ears and throat as well as conveying sensations from these parts back to the brain. Memory, vision, speech, muscle movements in particular parts of the body, coordination, balance, and other functions are controlled in various areas of the brain.

Memory, recognition, and personality are not localized to any specific cerebral area. Other functions have been traced to specific areas.

The hypothalamus and thalamus are at the core of the brain, atop the brainstem. The hypothalamus is an endocrine regulatory center that affects sleep, appetite, and sexual desire. The thalamus is a collection of nerves whose function is the integration and transmission of the many human sensations.

The spinal cord is the body's central communication network. It transmits signals to and from the brain to and from the farthest reaches of the peripheral nervous system.

Critical to the function of the brain is the arterial blood supply, carrying oxygen and nutrients. Though small in size and weight, the brain uses 20% of the heart's output of blood, 20% of the oxygen consumed by the body at rest, and 80% of the daily water intake. The blood is carried to the brain by arteries which extend through the neck from the heart, via the brainstem.

This system is complicated and vulnerable. Degeneration of nerve cells cause various physical illnesses.

A head injury may or may not cause a *visible* structural change. The exact type will depend on the nature, depth, location and degree of the injury. Head injured people may suffer physical, emotional and intellectual handicaps, while cognitive disorders cause problems in mental processing, such as focusing attention or interpreting written words, impeding learning.

Regardless of the source of the injury, the damaged area of the brain, to a large degree, determines the resulting symptoms. Though no part of the brain is isolated from another part, an injury may limit the activity of one part while allowing other parts to continue to function normally.

INTERPRETING THE PROBLEM

"The non-uniform inclusion criteria for mild CHI across neurological centers may account for inconsistencies in the extent of neurobehavioral recovery reported in various studies," according to Williams et al.

According to Berrol, a report from a committee of the American Congress of Rehabilitation Medicine indicated that " mild head injury (MHI), by definition, should not include an injury with a post-traumatic amnesia (PTA) exceeding 24 hours."

According to Thomas Kay, patients appear fine after a "minor" head trauma, until they attempt to resume their responsibilities at home, work, or school...a significant number experience great difficulty...complaining of inability to remember, concentrate, organize, handle more than one task at once...get as much work done as efficiently as they did before the injury...relations with family, peers and bosses often suffered, inevitably resulting in psychological problems. Their doctors were unable to find anything wrong with them, and they were seen as having psychiatric problems -- or worse yet, to be malingering. They became the bane of neurologists, psychologists, psychiatrists, and vocational counselors, whose usual techniques did not produce positive results.

There are two types of "mild" head injury - DIFFUSE MHI and FOCAL MHI. Though they differ in degree and results, "the practical result is the same ... Quick medical recovery followed by discharge home, with no formal plan for rehabilitation," according to Kay.

Of these two distinct types, diffuse and focal, I had the diffuse MHI. In the diffuse injury, the symptoms manifest themselves in numerous ways: the speed and capacity to process information; shifting attention to and from more than one task at a time; learning; memory; flexibility of thinking; complex problem solving; creative and abstract thinking; difficulty in expressing thoughts concisely; discriminating between situations; and capabilities normally taken for granted.

In the focal injury, a single, specific part of the brain is affected. The symptoms may involve attention and concentration; planning and reorganizing; learning and memory, and emotional control.

In the focal, changes are confined to one brain site, but "diffuse lesions are spread out and involve several areas." Billions of brain cells in these areas are distorted by the twisting, stretching, and compressing forces unleashed during the injury. Physicians may use such terms as "mild concussion," "concussion," or "diffuse axonal injury" to describe this injury.

"In a mild concussion, a person experiences temporary neurological problems but does not lose consciousness. ...A person who has a full concussion loses consciousness for a period of time and loses the memory of events just before and after the impact. Many people have no lasting ill effects from a concussion; however, some may show permanent subtle changes in personality and more prolonged memory loss.

"In both the mild and severe concussion, the regions in the base of the brain involved in breathing and heart rate -- called the brain stem -- are temporarily disturbed. The parts of the brain that control memory are also affected." NIH Publication #804-2478 (1984-Page 9)

"Often, these two types of damage occur together and produce overlapping results; a concussion with temporary loss of consciousness by some bruising on the front-temporal area. This is a classic closed head injury, occurring most often in moderate to severe injuries. In its mildest form, however, patients may appear quite "normal" and be discharged directly home." (Kay, 1986)

Old learning is intact. Deficits in learning and memory and short term memory loss are specific to new information, limiting the spontaneous recall of newly learned information.

For the executive or individual with a complicated job, deficits in planning and organizing limit the ability to plan, organize, initiate, monitor, and adjust thinking and behavior.

There is a third type of diffuse lesion, DIFFUSE AXONAL injury. It involves a damaged brain stem as well as torn brain axons, the fiber-like projects from nerve cells that help transmit chemical "messages" from the brain to the body.

Diffuse axonal injuries range from mild to severe. All forms involve coma, a loss of consciousness triggered by brain stem damage. Recovery of consciousness is followed by confusion and memory lapses. Some patients recover to resume normal activities, but others suffer intellectual, memory and personality losses.

These multiple symptoms "are organically based...caused directly by damage to nerve cells as a consequence of trauma." They are not psychological reactions to the injury or stress, as I was led to believe. This does not mean that secondary psychological consequences do not occur; they certainly do. However, emotional and behavioral problems can occur directly as a result of the injury, and it is essential (although often extremely difficult in practice) to distinguish one from the other.

Kay's observation, "This distinction is0020crucial because secondary psychological reactions may be amenable to more traditional psychotherapeutic treatment, organically based problems are not," makes an important point, which, when not applied, is not easily rectified.

When discharged, Kay states, "...the patient is unprepared for the difficulties they will encounter because they have been (implicitly or explicitly) misled into a set of expectations that exacerbate the problems they will encounter. They can walk and talk, dress and feed themselves, show no residual neurological abnormalities, are oriented, able to answer questions, and pass a mental screening test."

"Although headaches are considered the most frequent posttraumatic symptoms, cognitive difficulties may follow closely behind and are frequently noted in individuals with posttraumatic headache (PTH). The ability to concentrate and remember important details is often a primary factor in determining an individual's ability to function in society, particularly in employment settings. Cognitive deficits are most evident when an individual is under a great deal of stress, overtired, or trying to do more than one thing at a time. Some of our patients were only aware of cognitive difficulty under these circumstances. This may be the case in employment settings. Patients who suffer from cognitive disturbances may only become aware of these symptoms after they return to work or increase their activity level. Some may have to assume occupations with lesser responsibility than before the injury...We have been impressed with the number of PTH patients who also have troubling cognitive symptoms that have been overlooked by other clinicians or even by the patients themselves." (Packard, et al 1993)

There is rarely, if ever, an in-depth neuropsychological examination that would reveal deficits. Compounding this is the fact that deficits do not manifest themselves immediately after the injury.

DIAGNOSIS AND PROGNOSIS

Neurologists and neurosurgeons treat nervous system disorders. Using a routine neurological examination consists of the testing of tendon reflexes, Patellar reflex (tapping of the knee joint), muscle strength, muscle tone, and sensory function as a means of diagnosing a closed head injury would rarely be productive especially immediately after a blow to the head. Signs and symptoms applicable to closed head injuries can be numerous and diverse; but diagnosing and pinpointing it is often difficult.

"Your physician also may ask you questions that help determine whether your thinking, judgment, or memory is disturbed. Difficulties also may be encountered in distinguishing between neurological and psychiatric disease. A careful assessment of the character and pattern of symptoms over time, in addition to laboratory tests, may be required to decide among several possible diagnoses." (Mayo Clinic Family Health Book, 1990).

"A concussion (is) a blow to the head producing momentary loss of consciousness, or something close to it, without detectable anatomic damage to the brain. Although researchers have investigated concussion, no one is entirely clear as to the nature of the injury involved. It appears there is some damage to the white matter of the brain, where long-distance traffic is carried through bundles to nerve fibers known as axons. Such damage can evidently occur whether or not consciousness is lost, and it leads to post concussion syndrome." (Wolpow, M.D. 1991).

To establish base lines of the injury, the level of consciousness should be measured and recorded in terms that will be understood by anyone. Length of time for which there was a continuous loss of memory -- Post-Traumatic Amnesia (PTA) should be estimated. "Islands of memory" before continuous memory returns, should be ignored. With cerebral concussion, and by definition of the Glasgow Coma Scale, there will be a loss of memory for the event. PTA is a valuable index of the degree of cerebral injury, and helpful information, especially several months after the injury. In my case, if this information if it was in the hospital records, no ever mention it or ask me questions that might have been a clue to how long I was unconscious. I found myself constantly astounded when I finally learned this. I seem to relive that astonishment each time I come across it again.

"Some guide to prognosis is provided by the mental status, since loss of consciousness for more than 1 or 2 minutes implies a worse prognosis than otherwise. The degree of retrograde and post-traumatic amnesia provides an indication of the severity of injury and thus of prognosis. Absence of skull fracture does not exclude the possibility of severe head injury. ...a head injury, often trivial, precedes the onset of symptoms by several weeks...usually with mental changes such as slowness, drowsiness, headache, confusion, memory disturbance, personality changes, and even dementia." (White, et al 1992).

Illness does not develop out of thin air, and mine certainly did not. Neither do symptoms after a head injury. Since proof of visible damage after a closed head injury is difficult to confirm, symptoms are the body's only way to alert the individual to the fact that a physical problem exists. I had more than enough clues to have alerted someone. (See list of symptoms at the end of this chapter.)

"There is no such thing as a typical brain injury. The effect of the brain damage varies according to the location and severity of the injury." (NIH Publication 84-2478 Aug. 1984)

THE IMPORTANCE OF CLUES

Clues to the degree of the severity after blow to the head, often do not manifest themselves at the time of the injury. But, if the patient was unconscious, should there be a question about whether or not there has been a concussion or should the only question be the degree of the injury...a question that often cannot be answer at the time if it is a Closed Head Injury. With no detectable "anatomic damage", and no visible demonstrable permanent physical impairment to the brain, the inability to ascertain the degree of injury, a medical conclusion of "mild" concussion is understandable. This term is rarely mentioned to the patient. When symptoms do begin to appear, the patient is assured they will "go away" within a short period of time. Unaware of the complexity of the problem, it is easy, therefore, for the patient to accept this diagnosis and prognosis with no questions asked.

Why then, when the primary symptoms do not disappear and more complicated ones do develop, the conclusion may be changed to post-concussion syndrome (PCS)...a term substituted for a more serious injury but implies a psychological rather than a physical one. Since PCS is not a diagnosis per se but a "title" for the cause of an underlying disorder, are these new symptoms an indication that a serious head injury has occurred that requires more serious treatment?

Because a CHI, or MBI, or hidden injury, represents a constellation of non-specific symptoms, with no identifiable or provable pathological basis, it too often implies a "subjective" reaction to the effects of the injury rather than the reason for real symptoms.

"Some studies have suggested the need for a subdivision into an early post concussion syndrome and a later or persistent post concussion syndrome when symptoms and signs persist for more than six months." (Evans, M.D.)

"There are a number of important factors to consider in making an accurate diagnosis of Mild Brain Injury (MBI or PCS). One of the more important factors...is the pattern of symptom recognition. The individual who can walk and talk well, but has difficulty managing emotions requires very different skills from the professional than the individual who has major limitations in physical and behavior management skills...the symptoms of MBI are rarely identified until the person has experienced a number of relationship problems, (sometime) after the injury. We generally see people get into trouble at work or school or they have problems with the people they live with." (Willer, Ph.D., 1996)

In an ideologically, confused, polarized environment, characterized by doubt, derision, and debate, and subtle implications, sufferers are forced to seek medical help and treatment wherever they can get it. On this searching-for-answers winding road, they become enmeshed, as I did, in the medico-legal system, with its conflicting opinions and treatments. Symptoms become chronic, employment problems increase, families become disrupted, and bewilderment and despair become a way of life.

Randolph W. Evans, M.D. (1994) states: "In recent years, the term "post concussion (or post concussive or post concussion) syndrome has become the most frequently used despite its apparent inadequacies. It refers to a heterogeneous patient population with varying degrees of injury to the brain and head. Individual patient characteristics, including age, gender, personality profile, education level, intelligence quotient, occupation, prior head injury, and drug and alcohol abuse, may alter the expression of the injury. When evaluating the individual patients, the physician should give each symptom and sign a cause or a classification, when that is appropriate, as specifically as possible."

How these late developing symptoms are interpreted makes the difference between the care and treatment the patient or victim receives. The view and approach to the problem is an important element in the patient's effort to return to the previous level of capability. This effort is helped if the view, or at least some consideration, is given to the fact that the symptoms are, or could be, caused by the concussion. It is hampered if the interpretation is there is no reason for the symptoms, because that usually contains the implication that the patient is responsible for causing them rather than that they are being caused by the concussion itself. Before six months had elapsed, I was well aware that my doctors were believers in this misconception.

According to Thomas Kay, until recently, concussions were thought to be purely transient in nature, akin to "short circuiting," with no permanent damage to nerve cells in the brain. Autopsy studies, however, have demonstrated that even a minor blow to the head, leading to only brief loss of consciousness and apparently complete neurological recovery, can result in stretching and tearing of nerve fibers throughout the brain, causing disruptions that can only be seen microscopically.

"Complicating the process of problem identification is the question of what kind of valuation will actually shed light on the nature of the problem. CAT SCANS, (Pet scans, MRI's, EEG's), or thorough neurological exams may fail to turn up any shred of neurological evidence, even when there is a legitimate organic basis to the subjective complaints." (Kay, Ph.D. 1986)

Today, in the early twenty first century, we still do not have the medical equipment to register brain alterations immediately after a concussion. Even after symptoms do develop, it is extremely difficult if not impossible, since the medical equipment to register the microscopic changes that cause the symptoms is still not available.

"Unfortunately, there is as yet no scientifically documented connection between identified abnormalities and posttraumatic symptoms in mild head trauma patients. However, current technology is not sufficient to evaluate all suggested etiologies. For example, diffuse axonal injury, a condition that has been detected on autopsy in severe fatal closed head injuries, would not be revealed on an MRI (magnetic resonance imaging). For the present, the relevance of imaging studies in CPTH (chronic posttraumatic headache) is primarily as a means of ruling out less subtle neurological injury (e.g., hemorrhages, hematomas); more sophisticated imaging techniques may soon change this". (Dukro & Chibnall, 1999)

"Paradoxically, individuals sustaining MHI (Mild Head Injury) often have more intense post-traumatic symptoms than those who sustain more severe head injuries. The persistence of these symptoms often result in feelings of distrust from the public, the legal profession, insurance companies, and even other physicians. Even in the face of legitimate post-traumatic symptoms, patients may still be labeled as "accident neurosis," "personality disorders" or "malingering" in order not to be able to receive compensation.

"New technological advances in experimental head injury and neuroimaging techniques have revealed that even mild head injuries (MHI) result in subtle changes in the structure and physiology of the brain." (Packard & Ham 1994)

"In evaluation of any physical problem with symptoms that are not proven objectively on diagnostic tests, the question of malingering or secondary gain often arises.

"Malingerers may be capable of faking unusual symptoms and a degree of impairment; however, it is much more difficult to fake subtle neurological symptoms in a consistent manner.

"Studies cited by Mendelsohn (1984) indicate that between 50 to 85% of head-injured workers fail to return to work after settlement...It seems clear that there is no longer any justification for a neurologist or a lawyer to stand up in court and affirm that it is well known that patients with such symptoms immediately return to work after their claim has been settled." (Hinnant, Ph.D.; Tollison, Ph.D. 1994)

Having been well aware of how difficult it was for me to dispel these views, and knowing full well how inaccurate and unfair they were, it was a relief to discover that through the use of a battery of neuropsychological tests, administered and interpreted by a skilled neuropsychologist, a good many mild traumatic brain injuries can be detected, including some instances even when hard neurophysiologic signs are not present.

LITERATURE AND CRITERIA

Appraisal of the existing literature on MBI is complicated not only by the fact that this injury does not lend itself to visible or physical proof, but because of substantial differences in the terminology used. There is no one accepted definition of terms, making it difficult to interpret the data, and lending it to "personal views" rather than medically accurate proof.

Jess F. Kraus, in Seminars in Neurology (Mar. 1994), Epidemiology of Mild Brain Injury (MBI) states: "Traumatic Brain Injury has been studied in great detail from a clinical perspective -- that is, hospitalized patients. ...Of particular concern is that the assessment of initial severity of the injury (and determination that the injury is "mild") is often the product of negative rather than positive clinical findings...and are thus missed by conventional case-finding strategies...Classifications of brain injury are often based on duration of loss of consciousness, and/or presence, or absence of skull fractures or neurological symptoms.

"...duration and depth of loss of consciousness are difficult to estimate in mildly brain-injured patients...when all other test results are negative the assumption is often made that the injury must be mild.

"Neurological problems may be present for months -- and possibly for years -- following MBI and are the product of an "interplay of organic and neurotic factors."

In spite of the volumes of available information on this subject, Kraus states that important information is still not accessible. The epidemiological literature on MHI, despite numerous clinical reports, remains fragmented and only minimally informative. Very few studies have provided detailed information...of the long- term outcomes of a representative group of persons with MHI...the studies lack uniformity of definition of "mild" injury. Further standardization of the measures utilized, the matching criteria, and the duration and timing of follow-up are needed and will improve the base of knowledge in this area.

In spite of the difficulties, can the symptoms from a concussion be confirmed as resulting from the injury? They can, but not easily. A portrait of the injured person...or a pre-history that can be documented...previous experience, life-style, and method of coping with problems is a major clue and a good place to start, as are measures of work-related behavior. Neuropsychiatric tests can help confirm or refute the validity of the patient's description of symptoms.

"The uninvested, unmotivated, indifferent worker rarely becomes, by virtue of brain injury, an eager worker. Sadly, the opposite is frequently the reality. The client who possessed a solid work ethic prior to injury is often prevented from similar accomplishment by injury-imposed restrictions." (Silver & Kay, 1989.)

It is sad that not having a visibly detectable head injury has the disadvantage of not getting proper medical care, and results in being accused of a form of dishonesty. This allows "ignoring" medically known problems to "subjectively" rationalize them away. The physician who finally diagnosed my injury did recognize it. The others did not, in spite of having the same information available to them to do so. Today, medically, this psychological "rationalizing" still exists.

The victim with a serious head injury receives all the medical care he needs. His injury cannot be ignored. This is understandable; the attitude toward less serious head injuries is not. Much less care and consideration is extended to the victim of a mild injury who wakes up in strange surroundings, bewildered by what has happened, yet is shortly expected to pull himself together, and even return to work, with the facile reassurance that nothing serious has happened to him. How much of this has changed?

The victims of a "mild concussion" accept that diagnosis, believes what they have been told and even return to work. It should therefore be no surprise when, while trying to do so, serious incapacitating symptoms develop. The symptoms, how they manifest themselves, how they limit that effort, are the only clues to indicate that a more serious injury has occurred than what the "victim" was led to believe.

In 1972, Bryan Jennett wrote: "the basis for the subjective physical and mental symptoms frequently encountered after mild injuries, the so-called post-concussion or post-traumatic syndrome, remain a matter of controversy."

"The whole field of persisting symptoms after head injury is full of paradoxes, with remarkable recoveries after severe injuries and prolonged complaints after trivial ones... Whilst there seems little prospect of reaching agreement about these criteria it is wise when comparing data from different sources to verify that there are no striking discrepancies.

"What should be taken to constitute 'a head injury' has never been defined.

"Whether or not a patient is judged to have sequelae depends very largely on whether he complains of them or whether the doctor specifically searches for evidence of persisting dysfunction. Some reports regard return to work as evidence of full recovery; others require a faultless run through a careful neurological examination and psychometric test battery."

Though he wrote this in 1972, it is interesting to note that, 22 years later, in Seminars in Neurology, March 1994, Jess F. Kraus, et al, wrote, there is "uncertainty about the essential characteristics of this condition despite new reports in the professional literature ... the appraisal of the existing literature on MBI is complicated by substantial differences among research investigations including (but not limited to) non-comparable definitions of MBI and non-parallel outcome measures."

Continuing to consider this injury a mild concussion rather than a possibly serious one adds stress, to those handicaps the individual already has, camouflaging the reasons for them to an even greater degree.

Rigid thinking continues to see these symptoms as "subjective." It treats them as part of a PCS, inferring they have no basis in organic pathology, but are caused by the patient's negative motivation and not the blow to the head. This is a medical impression, difficult to hide from the patient, adds to stress

We do not call a patient's reactions to cancer a post cancer syndrome. Why then do we label a patient's reactions after a head injury a PCS? These reactions are the result of a known insult to the body, though admittedly, the cancer is visible, head injury is not.

Jess F. Kraus states that the epidemiological literature on MHI, despite numerous clinical reports, remains fragmented and only minimally informative. Very few studies have provided detailed information...of the long term outcomes of a representative group of persons with MHI...the studies lack uniformity of definition of "mild" injury. Further standardization of the measures utilized, the matching criteria, and the duration and timing of follow-up are needed to improve the base of knowledge in this area.

In 1995, I was advised by the Department of Health and Human Services of the National Institute of Health that "no agency currently collects such statistics on the after effects of closed head injuries (CHI). All of the figures must be extrapolated from other data." The data analyst of Aetna Life and Casualty, in response to my request, wrote: "...unfortunately, we are unable to provide the statistical data you are looking for."

Just because how to heal or help control the symptoms that develop from a concussion have not been discovered as yet, is no reason to conclude the injury does not exist. Viewing this injury for what it is, a malfunctioning of the brain and not as a psychological response by the patient would be a first long step toward acknowledging this difficult to diagnose injury.

Just because the mechanism for healing or helping to control the symptoms that develop from a concussion has not been discovered as yet, is no reason to conclude that no injury exists. The mechanism and information for healing this complex injury will emerge when investigators continue to search for them.

Recovery in general is uneven and the ultimate state of those who develop these symptoms has been poorly documented. Today, the relevant question is not only how to cure this injury but, even more important, at this state of our knowledge, what can be done to alleviate it until a cure is found. Camouflaging the true reason for the symptoms by spreading the "unexpressed theory of neurotic and malingerer" like confetti at a wedding celebration, allows this misinterpretation to fall wherever it may regardless of the damage to the victim.

"There is no specific treatment for post concussion syndrome. (Nor is there any for mild concussion.) Reassurance and support are, however, critical. Patients need to know that they aren't crazy or malingering. Their symptoms are real. Meanwhile, physicians, employers, coaches, teachers, and family should respect the fact that the working of the brain can be seriously affected by even a bloodless blow to the head." (Wolpow, M.D. 1991).

It is difficult to understand how these attitudes persist when information about alleviating the problem is available in our medical libraries. However, it is important to remember that professionals are no different from other human beings, and a powerful ideology can shape our beliefs in spite of strong empirical evidence to the contrary.

YES, THE CAREGIVER CAN HELP

Being a fastidious, creative, and flexible observer, attuned to all aspects of the debilitating results of this injury, are basic requirements for the caregiver forced to deal with the patient with a CHI, an injury that requires attention and special knowledge if the patient is to achieve progress in overcoming the inevitable handicaps the symptoms will create.

When physically disabled persons can see their disability, rehabilitation encourages them to be hopeful, thus hastening the recuperative process and lessening the potential for the development of depression, anxiety, and stress.

Mild head injured persons do not see their disabilities, and their injury drastically alters their thinking process. Without knowledge, encouragement, and advice on how to "cope", they remain in the dark with no understanding about what is happening to them.

It is necessary for the caregiver, to be aware that everyone's method of functioning is different, to anticipate and recognize injury-imposed sequelae and its effect on the skills of the patient, to find a way to aid the patient in ameliorating and compensating for them. Unaware of the nature, consequences and interaction among dysfunctions (since none exists in isolation) the patient is unable to know what to do with their impact and the changes that will develop in their ability in what had not been a problem prior to the injury…situations like employment, personal relationships, and routine daily living requirements. The need to be helped to develop an informed understanding of the problems and how to live with it is essential. The caregiver can do this with the proper approach and information.

Despite the prognosis that there is no acknowledged reason for these problems, the injured person will experience significant cognitive, emotional and behavioral deficits that seriously interfered with the ability to lead the normal life, pre-injury.

Those patients with no demonstrable, visible, physical or cognitive impairments are rarely referred from acute care to rehabilitation facilities. Instead, they are sent home to recuperate. Then, in response to economic and environmental pressures, they embark upon ill-advised pursuits with devastating results.

When these secondary symptoms do begin to develop, they need attention, advice and care, immediately.

There is an urgent need for both professionals and the lay public, those who do the treating and those who need that care, to increase their awareness and develop a sensitivity to this difficult medical complex and often misunderstood area, in the hope that by doing so the hidden injury can be prevented from becoming a permanent PCS. By then, the difficulty of reversing the damage the injury has caused becomes an even more difficult and often, insurmountable, task.

A final essential part of treatment is the least expensive. The neurologist should provide education for all concerned; the patient, family, other physicians, employers, attorneys, and representatives of insurance companies. Partly because of the Hollywood head injury myth, (where action, detective, western, and boxing films portray quick recoveries from injuries that appear serious...and in slapstick sequences and cartoons head trauma is portrayed as funny) the public is often misinformed about the effects of MHI. A treatment program of education, short-term bed rest, and timely follow-up may hasten recovery in some patients. Perhaps better education of society as a whole could lead to support for higher levels of funding for basic and clinical research into all aspects of head injury so that in the next century we will have more consensus and less controversy.

It was this chapter that brought with it the realization that I had developed a distasteful trait: the inability to tolerate improper actions In spite of the fact that this chapter did contain many explanations to my many questions, it also led to more questions: Why so many patient/physician communication problems? What are the roles of attorneys and insurance companies? What role do attitudes play in this drama? How and why do they develop? What process gives them such long life?

The answers I found were disturbing. They were an indication of how complicated and intertwined is the road that leads to procedures, based on beliefs that are so readily accepted, leading to wrong conclusions.

Symptoms of traumatic brain Injury

Initial symptoms that can be seen shortly after a head injury

Change in pupils – either too large/small or unequal in size
Convulsions -
Fluid drainage - nose, mouth, or ears (may be clear or bloody)
Vomiting/nausea
Bruising and/or swelling at the site of the injury, or flesh wound (external wound)
Low breathing rate or drop in blood pressure
Vision impairment – blurred vision, seeing double, vertigo
Difficulty tolerating bright lights
Inability to move limbs (one or more)
Loss of consciousness, or drowsiness
Dizziness
Slurred speech
Confusion
Amnesia

Symptoms that happen over a longer period of time

Initial symptoms linger on after visible wounds are healed
Symptoms improve, and then suddenly get worse
Impaired hearing
Smell – inability to smell, or a heighten sense of odors
Taste – lack of taste or change in how things taste
Irritability (especially in children) -
Personality changes - changes are often an exaggeration of the person's pre-injury personality
Unusual behaviors – aggression, depression or temper tantrums
Confusion - feeling disoriented, difficulty remembering, and making decisions.
Short term memory problems
Restlessness – inability to relax
Attention deficit – inability to concentrate or loose train of thought
Clumsiness, or lack of coordination
Insomnia – having an inability to fall asleep and stay asleep
Stiff neck – muscle tightness that leads to painful movement
Headaches – migraine or dull ache
Pain in head – different than a "headaches," this could be a pin-pointed or piercing pain.
Difficulty controlling or feeling emotions
Lack of emotional responses - such as smiling, laughing, crying, anger, or enthusiasm or their responses may be inappropriate

CHAPTER 10

Digging Beneath the Surface

> The post-concussion and whiplash syndromes have been controversial topics among physicians for many decades...Sizable minorities of physicians believe that once litigation is settled, symptoms quickly resolve. The available evidence does not support bias against patients because they have a compensation or litigation pending.
>
> -- Randolph W. Evans, M.D. et al (1994)

In my search for answers, I uncovered information that had been written on this subject throughout this century. In discussing this with lay people as well as physicians, I learned that patients were still being treated today as I was treated a quarter of a century ago. I was outraged that this important and readily available information was being so widely ignored. That outrage took the form of asking myself, "How could this STILL be happening?" I set myself the goal of finding answers. I think I did.

TBI has been studied in great detail from a clinical perspective...that is, in hospitalized patients...definitions of MBI vary widely. Of particular concern is that the assessment of initial severity of the injury (and determination that the injury is mild) is often the product of *negative rather than positive clinical findings*. Additionally, many cases of mild injury are either not witnessed, unreported, or treated on an outpatient basis, and are thus missed by conventional case history strategies.

Actor William Gargan once said, "Medical statistics are human beings with tears wiped off." I and so many others are part of those statistics.

Unaware of Kraus's words, for almost three years I lived through the undulating process of thinking I was getting relief, feeling better, and then sliding back. The true problem, exacerbated by this "zigzagging" process, prevented me from improving. In spite of being locked into this unrelieved pattern, separated from the normal flow of my life, submerged into the "illness" stage, I refused to give up searching for answers.

Though tears decreased, fears increased. I had to do something, even if only to shift my thoughts from where they were, on me, to something productive.

But what, where, and how to begin was not an easy decision to make. With so many unanswered questions, if there were answers, would I be able to find them?

The word "question" triggered an experience that took me back to my years in New York, during the depression of the 1930's, when my co-workers criticized me for "asking too many questions." I could not understand their objection then, but now, sixty years later, I have a greater understanding of their reaction. I was becoming annoyed with myself for that same reason. But the questions would not go away. I had no choice but to begin my own search for answers since I did not seem to be able to get them from anyone else.

Though I was fortunate enough to have eventually been diagnosed correctly, so much remained a mystery. Finding answers requires asking the right questions. Those that occurred to me were: How could I have changed so drastically? Had the information been available that could have prevented the end result? How could anyone think I was "neurotic" and a "malingerer"? Was I responsible in any way for that view? And if I was, why, and what could I have done to avoid those conclusions? If it was not me, what was it and how and why did it happen?

My search for answers resulted in this book.

Even before this project was born, I would read anything on the subject of concussion, hoping to find some clue to what could have caused so many changes, and what might "cure" some of my problems or even possibly help me in my effort to return to normal. I began haphazardly, in a most disorganized and superficial manner...reading newspaper stories, magazines, hospital bulletins. Never expecting to learn as much as I did, I kept no record of the sources...a major requirement in research. My primary motivation had been curiosity about whether what happened to me was routine and, hopefully, finding some informative material on what to do about it. I had no plans to do anything more productive with the material beyond helping myself. I certainly didn't think I would ever write a book.

But when I found what I did, facts that had seemed trivial and unimportant until then, now became relevant. Subtle bits of information emerged, forming a pattern. This expanded my interest to encompass not merely what applied to me but the broader subject of concussion, MHI, and the devastating damage being done to others like me.

As my knowledge increased, it only served to raise another question. I now began to search where I should have started in the first place... the public, medical and law libraries. My fascination with the subject was sparked by the realization that, in spite of the volumes of information on the subject, it confirmed the fact that victims of this injury were still being treated as I was...discharged to return home, told to resume their "normal life style," with no advice or information to enable them to do so...no more information than I had received so many years ago.

During the physical examination, special attention should be given to the level of consciousness and extent of brain stem dysfunction. My hospital record contained no information about how long I was unconscious or when I regained consciousness. Nor did it contain any information about whether there had been or could possibly have been a neck injury. Neither physicians, attorneys, or the insurance company checked for or were concerned about this lack of such relevant information.

Most TBI patients living with constant symptoms find themselves increasingly centered on the problem, while continually seeking some sort of relief. As patients go from doctor to doctor, with hope of answered questions find that hope dashed by more failure to find those answers. They often experience increasing depression and anxiety, often encountering subtle suggestions that their problems are "in their head."

This seemed like one clue to why my cognitive abilities, my greatest asset, to think clearly, to analyze and make decisions, my judgment and perception, prevented me from returning to normal. Additional confirmation was not long in coming.

E. Marcus Davis, Attorney at Law states the problem succinctly: "The injured person can learn to compensate for many, but not all, deficits rooted in permanent damage to brain tissue…(and) must receive rehabilitation that addresses cognitive deficits, emotional damage, and resulting behavioral problems. The objective of this treatment is the permanent resolution of the person's emotional distress as well as his or her reintegration into the community. According to one prominent neuropsychologist, omitting any of these components in treatment can lead to the ultimate failure of the treatment as a whole. Including all three can lead to a 'whole person treatment approach to which patients with minor brain injury respond favorably.' Cognitive remediation, as this type of therapy is called, combines cognitive psychology, neurology, remedial education, and psychotherapy. The therapist attempts to restore thinking and problem solving skills and to reteach social skills."

Knowing full well the problems these people have to face, the incentive to find answers became an even stronger motivating factor and, in the end, led to answers to my growing list of questions.

Though I never intended to include the role of the attorneys and insurance companies as part of my early research, it soon became obvious how important were their roles. What I found was no less devastating than the answers to the medical role in this controversy, which was slowly beginning to look like a fiasco.

Gradually, inexorably, what I learned created a host of wide-ranging conflicting emotions…leading from the depths of despair to finally being able to see the light. First came HORROR; horror upon learning of the volume of relevant knowledge that, had it been used, could have avoided those years of misery, misinformation, depression, doubts about myself, and the price my family paid for it. This lasted a while until ANGER slowly managed to escape from its hiding place, anger at the realization that had my voice been heard, the problem could have been medically alleviated from the very beginning.

With the help of my physician, and my strong desire to get well, as the light began to appear, a healthy response, APPRECIATION, made it possible for me to recognize my good fortune at having been diagnosed correctly, even if it was not until three years after the injury.

Sandwiched between these conflicting emotions is one that is probably responsible for my strong desire to write this book. It was my sense of DISBELIEF…disbelief that all this knowledge about what could and should have been done was at the fingertips of the physicians who misdiagnosed my symptoms. The question still lingers, were they really not aware of what I had learned, or had they simply ignored it, and if so, how could they and why?

The enormous volume of medical literature was an indication of the depth of the controversy that surrounds this subject. Were the divergent views of a concussion, whether labeled mild, hidden, or Closed Head Injury (CHI), a recent phenomena? Not so, these conditions were described at least a few hundred years ago as cerebral concussion with persistent symptoms in 1827 by Astley Cooper, in 1882 by Boyer, in 1889 by Dupuytren."

John Erichsen, Professor of Surgery at University College Hospital, London and surgeon extraordinaire to the Queen, in a total of 14 lectures in the last quarter of the 1800s, coined the labels "railway spine" or "railway brain" injury because he attributed it to "shocks of the body received in collisions on railways" and "obscure injuries to the nervous system."

Sixteen years later, in a revised treatise that included 53 patients who had sustained head injuries, only 17 were in railway accidents. Thirty-six were from falls or blows to the head. Nevertheless, aware of the controversy this engendered, he wrote: "...there is no class of cases in which more discrepancy of surgical opinion may be elicited."

In 1883, disagreement was voiced by a London surgeon, Herbert Page, "Erichsen was scientifically inaccurate, and suggested "general nervous shock," and "functional disorders" as his explanation."

I could not help wondering if "functional" had the same psychological implication then that it often has today in the medical vernacular? Those who accepted the theory that the patient exaggerated the symptoms in the hope of compensation, coined another label, "traumatic neurosis." Other terms offered were "vasomotor symptoms complex" for post-traumatic cases due to "disordered intracranial circulation."

It is interesting to note that this controversy about concussions with "invisible" injuries took the same path in the United States as it did in Europe, except that we now had automobiles, and the term "accident neurosis" was coined. No one questioned how it was possible that victims of concussions in the United States miraculously developed the same symptoms as their counterparts did in Europe, were viewed with the same distrust, and subtly accused of being neurotic.

Though the controversy has barely changed, the act of renaming the injury continues. Post-concussion syndrome, post-concussion amnesia, traumatic amnesia, mild traumatic brain injury, MHI are but a few of the diagnoses used to describe the symptoms that develop after a concussion.

In 1883, the Boston Medical and Surgical Journal summarized the controversy in an editorial:

"...It is natural that the medical 'bugaboo' raised by Mr. Erichsen some years ago, should meet with little quarter at the hands of the modern scientific observer...it is possible the skeptic may have gone too far...there may actually have been some phosphorescent light which we do not understand, and the nature of which we cannot fully explain...the cases recently reported in the Journal...point to the reality of a set of symptoms induced by traumatism, which correspond with those hitherto termed spinal concussion, a name so misleading that many accurate observers through the influence of the name alone have been induced to deny the existence of what the name covers. "

Russell C. Packard (1992) has said that H. Miller's article on accident neuroses was very influential, having reported that patients' nervous symptoms (not just PTH - post-traumatic headache) resolved for the most part following settlement. His description of a "typical patient" outlines several features noted in patients with some chronic pain who are often not destined to recover.

Symonds took an equally strong opposing position in 1962 when he wrote: "It is questionable whether the effects of concussion, however slight, are ever completely reversible."

How could it be, that, as late as 1961, in contemporary courtrooms, defense attorneys are (were) quoting from the writing of Miller, who summarized the viewpoint of those who believe that the post-concussion syndrome is really a compensation neurosis, as "The most consistent clinical feature is the subject's unshakable conviction of unfitness for work..." Physicians still harbor his prejudices even though Miller's position has been discredited.

Though there is important data to refute these conclusions, both beg the question. Is the only alternative to "accuse" the victim of responsibility for the symptoms that develop?

That this attitude still prevails today in spite of the following, made it more and more difficult to understand. "The post-concussion syndrome follows usually MHI and comprises one or more of the following symptoms and signs: headaches, dizziness, vertigo, tinnitus, hearing loss, blurred vision, diplopia, convergence insufficiency, light and noise sensitivity, diminished taste and smell, irritability, anxiety, depression, personality change, fatigue, sleep disturbance, decreased libido, decreased appetite, memory dysfunction, impaired concentration and attention, slowing of reaction time, and slowing of information processing speed...Loss of consciousness does not have to occur for the post-concussion syndrome to develop." The definition of concussion varies in different medical dictionaries, with some requiring loss of consciousness and others not. The origin of the word "concussus" is the past participle of the Latin verb concutere, "to shake violently." The term "commotio cerebri" introduced by Pare in the 16th century, has a similar meaning. Another symptom that may develop, but rarely discussed in this context, is epilepsy.

Admittedly, diagnosing this injury correctly is not a simple matter. The symptoms develop slowly, over time, in the progression they do, depending on the patient, the lifestyle, the work ethic and requirements. These symptoms bring on physical, psychosocial, and cognitive changes, with the cognitive becoming obvious only after the victim tries to return to the normal lifestyle practiced before the injury.

Post-concussion syndrome is used to explain such symptoms. At which point is this term applied; when there are two, five or fifteen symptoms? And what does it mean other than that the symptoms developed after a concussion. Does it mean the same to every physician? Or is post-concussion syndrome a valiant effort to squeeze a jumble of these disparate, and eventually chronic and difficult-to-live-with symptoms, into a neat package eliminating the need for a more accurately defined diagnosis, implying "the problem is all in your head?" That the brain is involved in signalling symptoms should never be more than a statement of the obvious.

Alexander J. Nemeth (1993) states the reason for this dilemma is - The front-line medical practitioner has been said to be notoriously delinquent in under-diagnosing the disorder. Pertinent information that has been accumulating in the journal literature for decades has not found its way yet to medical reference sources, or even neurological textbooks, with few exceptions. The readily available reference literature generally subsumes the head-trauma aftereffects under the ill-defined phenomenon of the PCS, described as a mélange of seemingly unrelated complaints, with no attempt to organize them into behaviorally meaningful categories. Yet, the symptoms and signs observed after head trauma lend themselves to be ordered; cognitive impairment, personality change, somatic complaints, and psychological reactions to the altered neurological state…neurologist's diagnosis of post concussion syndrome has no precise diagnostic meaning. The medical reference in literature is inconsistent in its use and definition of the syndrome. The syndrome is usually applied to a mixed group of trauma symptoms, for which there is no apparent explanation and is thus seen as a transitory phenomenon of no significant consequence…The subject is surrounded by myths that are generally accepted by the medical and legal professions, who solemnly enunciate them without recourse to scientific observation or authoritative studies.

Although extensive data of the last two decades strongly support an organic basis for the post-concussion syndrome, much doubt still exists among some physicians as well as laypersons, defense attorneys, and agents of insurance companies. Finding an acceptable answer to how and why these doubts still exist eludes me.

Randolph W. Evans, et al, found that very little "formal" medical information is available on the opinions and practices of physicians on post-concussion syndrome and whiplash, despite the acknowledged controversy. The fact that only one opinion survey (1967) has been administered to neurosurgeons on post concussion syndrome and none on whiplash is one indication of the indifference to this serious problem. To fill this gap, in 1992 Evans performed a national survey by mail of four physicians groups commonly treating these problems... family physicians (FP), neurologists (N), neurosurgeons (NS), and orthopedists (O).

Evans notes that the term *whiplash*, first used in 1928, is a term that covers "typical hypertension followed by flexion of the neck...and commonly occurs after side or front impact collisions, rear end collisions and are responsible for 85% of all whiplash injuries...(which) in addition to neck pain, can also include back pain, headaches, dizziness, paresthesias of the upper extremities, and psychological and somatic symptoms." Today, *whiplash* a clearly descriptive diagnosis, has been eliminated as a diagnosis and *soft tissue injury* (also called acceleration/deceleration injury) has been substituted, as a less descriptively specific term, less understood, seemingly less serious, and more readily ignored by the victim. Those professionals who do not view this neck injury as serious believe that whiplash encourages the victim to psychologically indicate "imaginary pains and problems." One must wonder if soft tissue injury will receive more serious attention than whiplash.

Mayou, R. and Bryant, B. (1996) found that there is little quantitative evidence for the belief that psychological factors are important in the course and outcome of whiplash neck injuries and [are] not associated with any baseline psychological variable or with compensation proceedings. However, without improved medical imaging techniques that will show specific injuries, many will dispute the existence of whiplash or organic brain syndrome. Until then, more appropriate treatment is required. Reduction in ensuing disability can be helped by early acknowledgement of the injury. Advice on how to cope with the symptoms will help reduce ensuing disabilities.

Prescribing specific treatment for the numerous symptoms that develop from a concussion, post-concussion syndrome, or whiplash is not an easy matter, and requires special training and knowledge.

The responses to Evans' survey created another unanswered question. Is it really possible that, in spite of the pertinent information available about this debilitating and destructive injury, physicians can find logical reasons for retaining these views?

Many aspects of PCS and WS (whiplash) are controversial among treating physicians. This controversy can have a profound impact on the quality and cost of patient care. Only a minority of respondents [to Evans' survey] believe that PCS and WS are clearly defined syndromes and believe that effective treatments are available. A substantial minority report that psychogenic and litigation factors are most responsible for the conditions. Most believe that PCS and WS have a 3-6 month recovery time and a significant minority (of physicians) concur that symptoms of the two syndromes resolve when litigation is settled.

Alexander J. Nemeth, (1991) in presenting a brief history of the problem in his article entitled *Blind Spots in the Diagnosis and Management of Minor Brain Trauma*, opens that section with:

"For the medical practitioner, undoubtedly, the single, most confusing aspect of posthead-trauma syndrome is its etiologically hybrid nature. This presents the diagnostician with a pseudo-dilemma: Is the condition organic or psychologic in its origin? It is, in fact, a complex composite of the two: There is the physical insult to the brain (mostly microscopic lesions in the white matter and brain stem) and there is the psychological reactions to the sudden disruption in the regular flow of neuronal and emotional/behavior functioning. To the unbiased student of the syndrome, the organic and psychological effects of head trauma are integral parts of one and the same condition and have to be understood in their interaction both with each other, and with the individual's background and unique personality makeup."

Physician and public perceptions may be influenced by popular misconceptions. As previously stated, the "Hollywood Head Injury Myth" can be traced to the depiction of head injuries on television and in motion pictures. This type of mythology does not die easily. This may be believable in the public mind, but how is it possible in the medical mind?

"...the available evidence does not support the bias against patients just because they have a compensation claim or litigation pending. Such bias can result in a patient becoming a victim of the injury and then a victim again of our adversarial judicial process.

"This negative bias on the part of physicians can adversely affect patient care and can compel patients to seek multiple physician opinions. Since 34% of the adult population in the United States uses unconventional therapy, many patients with PCS and WS seek out alternative treatments as well." Evans, et al (1994).

In fact, one neurologist who responded to the Evans' survey, surprisingly, courageously stated: "I am pleased to see the diversity of treatment options in the questions. It's about time we as MD's wake up to other alternatives within our referral power. I refer both to acupuncturists and chiropractors not because my patients ask for referral, as they often do, but because of the positive results that they obtain."

After a concussion, statements made, again and again, are... there often are no physical signs to suggest a more serious injury; the injury appears slight, consciousness usually returns promptly; the patient appears to recover with no lasting ill-effect; there may be a short period of disability afterwards, with numerous and varied symptoms. Though there are numerous symptoms, as previously mentioned, only the less serious but obvious ones, memory loss, headache, irritability are usually recorded, and the patient is discharged.

Nevertheless, concussion is not necessarily a minor injury. There may be signs that more serious symptoms are slowly developing. These signs often are an indication the concussion has done more than "mild" damage as originally diagnosed.

Serious brain damage may follow blows to the head, either with or without fractures of the skull. Concussion, indicated by loss of consciousness but with no "detectable" signs to indicate a "serious" injury is no indication that there is none. The longer the patient remains unconscious, the more severe the injury is likely to be. Is ignoring this statement sound medical thinking?

Behavior following injury is also influenced by a host of other factors. Environmental factors, stage of development, effects of normal grieving, personality characteristics prior to the injury, and personality disturbance that may develop after the injury due to coping or adjustment problems contribute to the person's behavioral pattern (Hibbard et al., 2000; Prigatano, 1999). The Mild Traumatic Brain Injury Subcommittee of the Head Injury Interdisciplinary Special Interest Group of the American Congress of Rehabilitation Medicine indicated that mild head injury by definition should not include an injury with a PTA (post traumatic amnesia) exceeding 24 hours.

Specific signs have a high likelihood of being related to a specific disorder. Non specific signs have a low likelihood of being related to a certain disorder. Many of the cognitive and affective signs of MHI are nonspecific. They are evident in patients with MHI and with other medical disorders, as well as in healthy persons.

"The most conservative assessment approaches must be used to maximize diagnostic accuracy for MHI. In the final analysis, diagnostic accuracy will elude the neuropsychologist (or physician) who does less." (Fisher, and Williams, 1994). The literature I found on this subject confirmed that is precisely what was being done, medically, and is still being done today.

"It cannot be disputed that every severe injury, defined as one followed by 24 hour post-traumatic amnesia, (PTA), has been associated with profound disturbance of normal mental activity in the early post-traumatic period. Accepting the current view that even brief unconsciousness is always associated with some degree of permanent change in the brain, it must be assumed that the physical substrate of mental function is always impaired to some extent after severe head injury; and indeed there are those who hold that some persisting alteration in mental function can always be detected if the search is assiduous enough. Unfortunately, the methods which are available for assessing mental function are very insensitive, even when applied to those features, such as intellectual performance or memory, which are to some extent measurable". (B. Jennett, 1976)

Since Bryan Jennett wrote this in 1976, it should be noted that great progress has been made in the methods "for assessing mental function" after a head injury. It is questionable whether attitudes have kept pace with this increased knowledge.

In an article in Lancet, a British prestigious medical journal as early as September 14, 1974, entitled "Delayed Recovery Of Intellectual Function After Minor Head Injury," an extremely important question was answered: Can the hidden injury be diagnosed before it has developed into a post-concussion syndrome?

Gronwall and Wrightson did a study that included patients who had been concussed with a PTA of less than 24 hours. Excluded were those whose symptoms were readily identifiable, including those whose headaches were due to vascular, skeleton or peripheral-nerve involvement.

Their study showed: "Patients who had suffered concussion are unable to process information at a normal rate. The time taken to recover is related to the severity of the injury, judged from the duration of post-traumatic amnesia, and after uncomplicated concussion is usually less than thirty-five days. Patients with post-concussion symptoms, who complain of inability to carry out normal work, poor concentration, fatigue, irritability, and headache, show a reduction in information-processing rate which is inappropriate to the time elapsed since injury and which persists beyond the usual period of thirty-five days. It is suggested that reduction of information-processing rate is an important factor in the genesis of the post-concussion syndrome."(D.Gronwall – 1974)

After concussion, two disturbances of intellectual function are evident. Post-traumatic syndrome or post-traumatic amnesia, terms often used interchangeably with post-concussion syndrome, depending on the physician, and/or the time span, are easily demonstrated and have been thoroughly documented. The second, attention and impaired speed of perception, is not so obvious.

Most patients who suffer a minor concussion recover rapidly and return to work. This is the group where the diagnosis of mild concussion applies accurately, and should be accepted medically. When the situation changes and additional symptoms develop, and the medical profession continues to accept mild concussion as a correct diagnosis, the problems become the problem.

After studying a control group of concussed patients who had recovered, they then studied information processing rates in patients who continued to complain of inability to carry out normal work, poor concentration, fatigue, irritability and headaches. The study group ranged in age from 17 to 55; had post-traumatic amnesia; some had returned to work but were unable to continue because of poor concentration, fatigue, irritability and headaches and had been diagnosed as having post-concussion syndrome. All satisfied the criteria for the study.

Though others have identified this syndrome as it is seen months or even years after the injury, Gronwall and Wrightson have identified it as a source of disability as early as a week after the accident. The subjective elements are accompanied by objective changes in intellectual function. They also found that as the intellectual function returns to normal, the symptoms regress.

The obvious and relevant question seems to be: would early attention to this injury, with proper care and advice, be reversible before it became residual? This "fixed liability" is the end result of the chain reaction of symptoms that begin to develop after a concussion and after the victim returns to normal activity. Though the symptoms eventually do abate in many patients, when they do not, as each new symptom is ignored, on the assumption it will "go away," it leads to other symptoms that lead to the end result, a post-concussion syndrome, a syndrome more complicated and more difficult to alleviate than the original symptoms of post-traumatic amnesia and intellectual disability when they first appear.

"The findings raise two important questions: the way in which the changes in intellectual function are related to symptoms, and the relation of early symptoms to the later development of post-concussion syndrome and accident neurosis."

To help recognize these early symptoms, Gronwall and Wrightson use the PASAT (Paced Auditory Serial Addition Test) as an index of the rate at which information can be processed by the concussed patient. It measures "channel capacity," "the amount of information that can be handled (by the brain) at one time, a quantity with definite limits. The processing of information by the mind will be inadequate either if the number of items demanding simultaneous attention is too great, or if the rate of function is inadequate." The concussed patient may be able to cope with numerous items if they can be handled sequentially rather than simultaneously. However, as the speed and number of items to be handled increases, the performance of these patients falls off. Their channel capacity has been exceeded and their rate of function limited.

Because so many symptoms are viewed as subjective, it is difficult to critique them against vigorous proof. However, in therapy Gronwall and Wrightson found confirmation of the theory of reduction of capacity impairing work performance, etc; that tasks can be completed without fatigue or complaint of poor concentration if the demands on capacity are reduced. If demands are increased, symptoms return.

It is important to note that personal attitudes and duties have a great deal to do with how disabling the symptoms might be. The responsibilities of the patient, after discharge, also color that response. If the demands are light, not work related; if there is no pressure, and rest is possible whenever fatigue is present; the disabilities are less incapacitating, and the patient sees them as less serious than they really are. Many patient may notice minor problems with memory or problem solving, but do not attribute these signs as seriously interfere with their life functioning normally, or that these minor problems require conscious adaptation.

If, on the other hand, the life style makes strong demands and there is a need to return to work for financial reasons, and rest is not possible when required, symptoms develop that limit the ability to function at the previous level of capability. The attorney cannot prosecute his case with the same success; the physician is handicapped in his ability to treat his patients; the salesman fails to persuade his clients; the parent cannot cope with the children,, the writer loses creativity, and the musician cannot perform. Executive abilities, once automatic, now require more time and thought before being capable of making a decision, no matter how minor. Loss of confidence and doubt about oneself develop. Anxiety follows.

"What differentiates this process from the anxiety of neurosis, is that it is grounded in an organically-based dysfunction...one cannot deal with anxiety without taking into account the very real dysfunction that fuels it." (Kay, 1989)

If resentment develops, it may manifest itself in the attitude toward those who do not acknowledge that there is a real disability...the physician, the establishment, the employer, the family.

"These are the patients the psychiatrists, psychologists, and rehabilitation people are asked to 'cure'. The organic origins of the syndrome are unlikely to be evident, having regressed long before in the natural course of events. At this point it may be impossible to reverse the chain of events and rehabilitate the patient; though payment of compensation may silence the patient's complaints, it will cure none save the malingerer." (Gronwall and Wrightson 1974)

CHAPTER 11

Medicine and Its Gatekeepers

There are no tricks...no shortcuts involved in improving, stabilizing, and salvaging the doctor-patient relationship...failure increases the likelihood of deadly medical mistakes and situations in which there are, ultimately and sadly, only victims.

--Charles B. Inlander, et al

"The art of medicine consists of three parts...the disease, the patient, and the physician. The physician is the servant of the art, and the patient (the recipient of that art), must combat the disease along with the physician." This statement has been attributed to Hippocrates, circa 400 B.C. and is still valid today. Just as the mind cannot be separated from the body, neither can one separate the patient and the problem from the physician and the art.

Medicine and its research are rich in accomplishments. Though medicine is an art, not a science, as some would like to believe, these accomplishments make fascinating reading while raising questions about the attitudes and commitments of some of its members to heal and care for their patients.

This complicated patient-physician relationship, especially in attitudes toward women, is colored by character, personality, and social factors as well as medical training and lack of "bedside manner." In the early 20th century, "bedside manner" was part of the medical curricula. As more and more mechanical techniques were developed, major contributions to medical progress, they replaced "bedside manners." Today, this training is being reintroduced into the medical training and is being stressed as an important factor in the health care field.

Doctors and patients come from all walks of life. Different backgrounds cause one to develop different ways of viewing the same situation. On the other hand, the relationship may or may not work under any circumstance. The physician may make the correct decision.

The patient, however, may be incapable of making a decision that should be easy, and instead, seems to be very difficult.

If the physician does not have the knowledge to make an accurate diagnosis, admitting it is something he may not wish to do. Whether this is because of ignorance, or because he does not want to destroy the patient's confidence in him and his ability, only he would know. On the other hand, he may be concerned about the patient's reaction to a clearly stated diagnosis and the effects of his explanation on the family members.

This relationship is fraught with handicaps. Inevitably there will be errors...human and otherwise. Patients and doctors are often strangers. Specialists rarely see their patients often enough to get to know them well. Busy physicians may "skimp" on personal case histories, missing the very clues a comprehensive lifestyle portrait may help provide.

Communication itself is a difficult art. It is all the more so because words frequently do not have the same meaning for the patient as they do for the physician. Patients may be too ill or too upset to understand what the doctor is saying or be able to remember instructions. When a prescription is written, the physician may not give the patient specific instructions about the best time to take it, or explain what possible secondary symptoms may develop. However Pharmacists are doing better at educating the patient on side effects and other drug interactions.

Myths about women are responsible for a great deal of biased treatment, and physicians with a problem determining where their bias lies find it difficult to discount these myths, even when challenged by reason. Specific attitudes toward women often affect the relationship.

Sorting facts from opinions is a difficult task for the patient. The self-interest of the opposing factions... patients, physicians, and, in some cases, attorneys and insurance representatives ...muddies the picture, each accusing the other of being responsible for the problem. None wants to take responsibility for their actions, thus adding to the "wreckage" with rhetoric. In this confusing atmosphere, communication and understanding become more difficult. Professionals and lay persons are caught in the crossfire of verbiage used by all parties to protect themselves.

There is a tremendous need for change in the doctor-patient relationship. Who is more likely to acknowledge and try to correct this need; doctors or patients?

As we learned firsthand, a kind bedside manner is not merely a quaint characteristic you hope for in a family doctor. A doctor's attitude toward a patient and the patient's family colors every moment of a health crisis.

It can help a patient to heal, keep those of us who suffer alongside her saner and healthier, and lower costs.

And yet for all the advances in medical technology and research, simple kindness from health-care providers is all too rare. A recent survey conducted by the Arnold P. Gold Foundation, which advocates for a respectful bedside manner, asked 600 people to describe their interactions with doctors. Twelve percent said they were taken care of by doctors who didn't know their names. Twenty percent had met with doctors they found "rude or condescending." Forty-seven percent said they had felt rushed by doctors. --Katherine Rosman, April 13, 2010.

This book is an effort to show that relevant questions that must be asked by both and, when answered, acted upon and not be allowed to fade away in the records. This book is not a defense of malingerers...there will always be some. It is, however, a call for professionals to devise a method to distinguish between the victim who is truly a victim and the malingerer who is a true malingerer. There must be some better or more reasonable method to make this distinction rather than the way subjective conclusions about this serious problem are reached today.

Though some people do invent symptoms, does this give anyone the right to conclude a head-injured victim is dishonest because physicians cannot find the physical proof for the symptoms when a known blow to the head is involved? Yes, some people do get better. But some do not. Is there any logic in concluding that those who do not are malingerers feigning their illness? The physician cannot prove there is no injury when the patient reports symptoms. Patients cannot prove there is an injury, in spite of the symptoms they know they have. Both have the same problem of believing what they believe on improvable subjective conclusions. What is the physicians' responsibility when they are faced with these problems?

My head injury and how it was handled is not important because it happened to me. It goes deeper than that. Though similar diagnostic problems and medical approaches to them occur in other medical disciplines, head-injured persons, almost a quarter-century later, are still being diagnosed as I was. Although post-concussion symptoms, once called "accident neurosis," were originally thought to be predominantly related to secondary gain, it is now clear that the prevalence of malingering was overestimated.

There are serious consequences to the lack of a proper diagnostic category for post-concussion disorder. First, communication among clinicians is made difficult by lack of precise diagnostic framework characterization. Second, the lack of a precise diagnostic framework has perpetuated the myth that many patients, especially those with 'milder' head injuries, exaggerate their complaints after the fact either because they are 'hysterical' or because they are motivated malingerers seeking some sort of compensation. Third, research on epidemiology, natural history, phenomenology, and treatment is hampered by the use of sometimes imprecise, incomplete, or contradictory schemes by different groups of authors. The overall result of these problems has been continued stigmatization of patients who are already suffering from neurobehavioral disorder occasioned by head injury.

The physician's problem is that he cannot find anatomical proof that there is an injury. The patient's problem is in knowing he has symptoms that cannot be seen by anyone else. The only visible changes are in behavior and persona, the very symptoms the physician views as "emotional" and "subjective," while ignoring the true cause, the head injury. Drugs may alleviate the symptoms, but only temporarily. They will also camouflage them, masking rather than solving the underlying problem. The patient's condition will continue to "worsen," disabilities increase, and new subtle changes in physical capabilities and personality will continue to develop.

A head injury, or head trauma syndrome, untreated or diagnosed incorrectly, creates stress which often leads to a diagnosis of post-traumatic stress disorder. Suggesting the patient see a psychiatrist adds another layer of distress to that which has already developed from the doubt and uncertainty about the reasons for the symptoms.

A complex syndrome is rarely either 100 percent organic or 100 percent psychogenic. Psychiatrists and psychologists are well aware that it is often a blend of both. Thus the post-traumatic stress disorder can be helped with psychiatric care, and often with good therapeutic results. However, just as a physician cannot cure a misdiagnosed illness with medication, so the psychiatrist cannot cure a physical injury with psychiatric advice. The danger is when a psychiatrist accepts the physician's diagnosis that nothing is wrong without questioning the basis for that conclusion. Believing there is no physical reason for the patient's symptoms, as the second neurologist and (insurance) orthopedist did in my case, the task for the psychiatrist becomes insurmountable.

If cancer generated as many early clues and physical changes in the body as a head injury generates from a concussion, more cancer patients would be treated immediately and saved. No physician would dare ignore "early" cancer symptoms, but some physicians ignore symptoms from a head injury. Claiming they are subjective, not real, results in a subjective "default" diagnosis.

Can a strictly psychological diagnosis that ignores a head injury be an objective one? Or could it be no more than a subjective bias on the part of the physician? A diagnosis not based on a thorough case history and relevant medical data could well fit this category. Patients have a right to expect physicians to curb "personal opinions." They may believe in them; they may live with them, but they do not have the right to allow them to influence their professional conclusions.

Nada L. Stotland, MD, MPH, comments: "The term psychosomatic, in common parlance and in the medical community, connotes physical symptoms that are not medically legitimate. At least two national organizations with the term in their titles have given serious consideration to changing it. It implies physical symptoms without organic concomitants, triggered by emotional conflict, exaggerated emotionality, stress, and/or a desire for attention or other personal benefit. It tends to be dismissive; psychosomatic illnesses are thought to consume inordinate amounts of medical attention and are not deemed worthy of that investment. Prejudices such as these belie the intrinsic interrelatedness of 'psyche' and 'soma', deters scientific investigation into a fascinating interface, and ultimately add to the waste of medical resources."

The original rise of the psychological paradigm within medicine was not communicated to patients. Instead, at the turn of the century psychologically oriented physicians continued to let patients believe they were receiving organic therapy, a viewpoint and attitude still prevalent today. Within ten days of a serious accident and concussion, the first neurologist's medical diagnosis that nothing was wrong is a perfect example of this attitude...

> *"I cannot help feeling embarrassed by my profession when I hear the myriad ways in which doctors convey their pessimism to patients. I would like to change this pattern and am working to require instruction in medical school about the power of words they speak to patients."*

> *"On rare occasions a medical "hex" may motivate an exceptional patient to prove the doctor wrong by getting well. The usual effect is despair, and I cannot believe that despair has beneficial effects on the human healing system."*

> *-- Andrew Weil, 1995.*

Attitudes of physicians, loaded with experience in what they think they know, while refusing to acknowledge or consider what they do not, are fertile ground for errors and misjudgments even if one does not questions motives. Bored VID's (Very Important Doctors -- or "top" doctors) who come to hasty conclusions about their patients, based on their experience and not on data, are like the aroma of a broken bottle of perfume, affecting all who come in contact with it. Professionals involved in these cases are influenced by these conclusions but rarely will acknowledge it. Even when they do, too often they do not, cannot, or will not speak up. It is rare to hear a doctor make an unflattering remark about another doctor, but they have no difficulty making them against their patients. Instead of a zero tolerance of the code of silence that permeates the medical profession, today many physicians live by their own Eleventh Commandment: "Thou shalt not speak ill of a fellow physician." Yet diagnoses made as indictments of a patient's integrity and mental stability are bandied about with no regard for their effect on the patient.

Intellectualizing a situation may encourage physicians to pay more attention to the similarities among their patients' illnesses rather than to the nuances that are the clues to the differences between them. Conclusions arrived at on the basis of similarities make that physician more like a "mechanical technician" rather than a professional humanitarian.

MEDICAL CARE FOR WOMEN

The basis for the difference in the way male and female patients are treated has its roots in our distant past, along with the myths we accept and believe for various illogical reasons. However, one does not have to go too far back to find evidence of it and some of its more recent roots in today's society.

With the advent of second-wave feminism during the 1970s, a significant body of literature emerged describing sexist practices in women's health care. Gender-biased diagnosing-the notion that somatic complaints by female medical patients are more likely to be labeled by physicians as psychosomatic-became a concern that garnered considerable attention in Europe and the United States because of the increased health risks it posed for women. This article examines the impact of feminist knowledge on this topic during the quarter century spanning 1970-1995. Analysis of the literature reveals feminist perspectives played a critical role in uncovering and problematizing gender bias in women's health care. -- Shari Munch

Though females average about 25 percent more visits to doctors than men, when sophisticated techniques are used to include women's longer lifespan and childbearing related visits, the disparity disappears.

People investigating why doctors treat women's complaint less seriously than men said that physicians think of women as malingerers. Physicians often assume that women's illnesses are psychogenic; that they originate in the mind, not in the body. With this assumption, it is not too difficult for physicians to prescribe mood altering drugs for women even when the drug is unwarranted.

It seems this medical attitude toward the difference in the sexes is not endemic to the United States. In the February, 1973, New England Journal of Medicine, Corea states that Dr. Jean Lennane, a psychiatrist, and Dr. John Lennane, Senior Registrar of the Renal Unit at Prince Henry Hospital in Australia, examined four female disorders, including dysmenorrhea (painful menstruation). In their study of nausea of pregnancy, they found that although 75 to 88 percent of pregnant women experience this condition, it is classified as a neurosis. Yet, the 1972 textbook Gynecology and Obstetrics, Current Diagnosis and Treatment, states that nausea "may indicate resentment, ambivalence and inadequacy in women ill prepared for motherhood", another indication that an "underlying sexual bias for this prejudice" exists. In 1980, 84 percent of drug using women in Canada reported that their physician had first recommended the drug to them.

Medical care for women, based on beliefs, half-truths, and mythology that they are basically hysterical and neurotic, has an illusory certainty about it that leads to fallacious thinking; a belief difficult for the patient to disprove and even more difficult for professionals to give up in spite of disconcerting evidence to the contrary.

> *"We conclude, from the research reviewed, that men and women appear to experience and respond to pain differently, but that determining whether this difference is due to biological versus psychosocial origins is difficult due to the complex, multicausal nature of the pain experience. Women are more likely to seek treatment for chronic pain, but are also more likely to be inadequately treated by health-care providers, who, at least initially, discount women's verbal pain reports and attribute more import to biological pain contributors than emotional or psychological pain contributors.*
>
> *-- Diane E. Hoffmann, Anita J. Tarzian*

Another report states the reason for physicians prescribing so many tranquilizers is, "Women ask for them," thus blaming the patient for actions no physician should be willing to admit. Would that same physician give chemotherapy to a patient because that patient believes that cancer is the problem?

A leading London psychiatrist told doctors (evidently British physicians have the same flaws), "the time had come when they should recognize that the patient who presented them with symptoms which could not be attributed to any ascertainable cause, should not be dismissed as neurotic and given a prescription for tranquilizers or anti-depressants..." This is not to say that tranquilizers should not be prescribed but rather that they should not be used indiscriminately...or to cover up the inability to find the reason for the patient's problem.

Physicians, impatient with not readily diagnosed symptoms, often are heard to say something along these lines, "People come in wanting sympathy, wanting a listener. A doctor absolutely should not be used in this way. Doctors are there to provide caring. Not listening."

This statement is dismaying. When there is no obvious reason for the patient's complaints; when the exact cause cannot be pinned down, as in my case and so many others, clues to the reasons for symptoms can often be found in the case history and background of the patient. The symptoms the patient tries to relate to the neurologist require the physician to listen. Listening requires time, which equates to lost money, and physicians seem to be complaining about it.

Could it be that patients' complaints about physicians "not listening" and "being intolerant of the patients' so-called complaints" is valid and a symbol, not so much of a complaining patient wanting sympathy but of the impatience of physicians who lack time?

Are physicians not susceptible to the same "foibles" and "neurotic" responses they often attribute to their patients? They excuse their practice of "defensive medicine" as a protection from possible future problems that might lead to a malpractice suit. Is this not a self-protecting mechanism? Is this the same, if not at least similar, to accusing a victim of being a malingerer? But more cases of malpractice are never filed vs. those that are. For those that are not, where can the patient go for redress? Does the medical profession have an open mind about the patient or do they tend to view the complaints as invalid? Could it be this attitude and not patient's greed is one of the major factors that has led to malpractice suits?

SECOND OPINION

When in doubt about the correctness of a diagnosis, don't accept it readily. When a physician "labels" your problem based on his beliefs rather than on medical data required to confirm his conclusion, beware. Seeking a second opinion is often wise, not just because some insurance companies require it. If they can doubt a physician's diagnosis, is not the patient entitled to do so too? But, it is urgent to continue to be cautious. Do not give the diagnosis or name of the first physician to the second one. His diagnosis should be based on his medical findings and not be colored by the conclusion of the previous physician. "Mental togetherness" is the act of professionals arriving at the same conclusion by accepting each other's diagnoses without ever even verbally discussing the subject. When the symptoms clearly indicate their cause, there is no problem and usually does not require a second opinion. When they do not, the result can be disastrous.

I made this mistake, and the diagnosis of my problem by that physician to whom I had gone for a second opinion is a perfect example of how serious a problem it can be. His examination and his diagnosis were mere recapitulations of what I had told him about the previous neurologist. His stress was on a minor symptom while he ignored all the relevant ones.

Another example is the orthopedist who examined me for the insurance company. I would like to suggest that this also be labeled a "second opinion". If this were done, then he too should not be allowed to see the patient's medical records until after he comes to his conclusion based on his examination, not on the diagnosis of previous physicians.

What do physicians do when they are confronted with a possible cause for a medical dilemma but with which they do not agree, or do not wish to incorporate into their thinking? In George Orwell's book, 1984, the walls of the Ministry of Truth contain slits for the disposal of waste paper on which is written the facts their regime considered unuseful. Of the slits, Orwell wrote: "For some reason they were nicknamed 'memory holes'." Do physicians who treat patients with a head injury for which, for whatever reason, they can find no physical proof, end up ignoring it, also have "memory holes?" Do these memory holes help them retain their belief in the myth that the symptoms of these patients are in their minds and could not possibly be caused by the accident?

When the complaints of these patients fall on deaf ears, they lead to tragic ends for the patient...a lifetime of living with incapacitating and limiting handicaps. The gigantic "coincidence" of so many victims, with a hidden head injury, all having almost identical symptoms should not be a difficult concept to accept. But it is. Instead of questioning why this is so, the symptoms are labeled with different names, all leading to almost same conclusion …if there is nothing physically provable then it must be in the "mind of the patient".

When the symptoms do not go away, as the physician assumes they will, they are then labeled a syndrome. The word "post" is added as a prefix to whatever title the physician gave them. Though the concussion was not considered important, was never mentioned as a problem; was completely ignored, the diagnosis now becomes "post-concussion syndrome". If it had been labeled trauma, it would have become "post-traumatic syndrome", or "post-traumatic stress disorder". The use of the word "post" indicates, of course, that it developed afterwards, with which no one can argue. What is never clarified is why, though the concussion was originally ignored, it is so readily acknowledged in the "post syndrome" stage. Its non-treatment in the early stages inevitably leads to this "post-development" stage. If the second physician recognizes this, then there is help in sight. In my case, it was not until the third neurologist did, by which time it was too late to prevent the symptoms from becoming chronic.

It is important to note that since a concussion or "closed-head injury" cannot be cured, even the proper treatment may not necessarily prevent the development of the symptoms that cause the "post-concussion" or "post-traumatic" syndrome. An important major difference to the patient's well being is that the injury is acknowledged as the reason for this syndrome and not the patient.

Why are there so many different names for a head injury, and one that so many physicians do not acknowledge exists? Most illnesses have a specific name…pneumonia, appendicitis, ileitis, and diabetes…names that do not confuse anyone and mean the same thing to the patient as they do to the physician. But the head injury we are discussing has "titles" that the patient, even if given "a title" more often than not does not understand what it means or how serious it might be. It does, however, allow physicians the freedom to interpret the injury according to their own bias.

Are so many "closed" or "not visible" head injured victims really neurotic and malingerers? Are so many physicians correct in this evaluation of the victims because finding proof of the injury is not within their grasp? Of course not. It would be a fallacy to assume either one of these statements is correct. Finding answers to this dichotomy would be a major accomplishment.

Often, once a judgment is made, virtually no amount of contrary evidence can turn it around. When information is ambiguous, people often leap to a wrong conclusion. To a large extent, they see what they expect to see, and reject any information that would challenge their already established point of view. Psychologist Rachel Hare-Mustin, formerly of Harvard, said: "Unconscious prevailing ideologies are like sand at a picnic. They get into everything."

Old myths and bad statistics do not die; they just rest comfortably in the minds of professionals as well as patients. Few people can discern their own conventional prejudices. They regard them as obviously true and dictated by "fact".

Perception is an active process in which people view the world with their own expectations. Patients keep trying to convince themselves they are going to get well, or they are getting well. Denial is used here as a form of self-preservation. Reality is overlooked or camouflaged. People do not so much believe what they see as much as see what they believe...or want to believe.

Errors about people and by people can easily slip by unnoticed as the mind buries this evidence under a carpet of the unconscious. Old errors can become perceptions that cast long shadows. Once perceptions are planted they are almost impossible to root out, become self-fulfilling prophecies, known to psychologists as "behavioral confirmation biases". Unfortunately, we all have them...patients as well as professionals.

And so, this book is a plea to the medical profession. Today, we still do not have the knowledge to cure this injury. There is a need to find a method to recognize and acknowledge that there may be one. There is a need to do whatever can be done to help the victim before the symptoms develop into serious problems and become chronic.

It is important to avoid concluding the reason for symptoms is psychological because finding physical evidence to prove there is one cannot be found. Those in the rehabilitation field of medical care are trying to "help" these patients but only after they have been discharged by their physicians.

The combination of talented physicians, research scientists, and inquisitive minds has helped medicine make giant strides that were inconceivable a mere half-century ago. Static attitudes carry the seeds of their own obsolescence. The ability to discern the true nature of these injuries, as well as to differentiate between the injured and the malingerer may well be within the grasp of the profession, although it is not aware of it yet. Until then, the medical profession must avoid placing the burden of this lack of knowledge on the shoulders of the victims, the very people least able to handle it. These victims have enough difficulty coping with the effects of their injuries without the medical profession helping to compound them.

CHAPTER 12

Attorney Involvement

You can hire the best tort lawyer in the land for the same fee as you would hire the least experienced. That fee is called a contingency fee.

-- Melvin M. Belli

PART ONE: The Client's Dilemma

After an automobile accident, the need to hire an attorney is often forced on the victim by the reluctance of insurance companies to accept the truth about the degree of incapacity resulting from the accident; such reluctance often requires adjudication and a court trial.

Selecting the right attorney for the job is a complicated and difficult task; in purchasing professional services, common sense is rarely a sufficient guideline. Professionals, by occupation and training, present themselves as capable of providing their particular services. Many do so competently, but others do not, and in some instances, the attorney may even be the main obstacle to a fair outcome. Professionals require considerable training, and it is difficult for a layman to judge the competence of the professional. But, the layman has the responsibility to do just this.

Nothing is easier to ignore than that which does not yield prompt solutions, and though there were many subtle danger signals after I hired my attorney, I ignored them. In my defense, I truly was too ill to act and ignorant of what to do.

The client's problems are numerous. Unfamiliar with the law, injured or facing financial problems, without experience or knowledge about how to cope with the insurance representative, the client feels forced to hire an attorney. The choice is usually based on someone else's opinion, such as a friend's advice.

In a legal situation, the client's responsibility is to try to understand the decisions the attorney is making and the effect they will have on the case. You soon learn you must live with the choices made. Unfortunately, too often the client depends entirely on the attorney's decisions or remains unaware of what is or is not being done or what should be done. Not knowing what questions to ask, how to interpret the answers when questions are asked, what to do when questions are not answered, whether to go along with the advice or insist on what she/he believes or thinks should be done, these are only a few of the dilemmas facing the client.

Injured and ill, making a sound judgment is often beyond the client's capability, because functioning at a normal, healthy level is beyond his or her reach. When a client has doubts about an attorney, suspicions develop. But suspicions, though depressing, are not certainties, at least not until they are confirmed. My experience with attorneys is just one example of how and why clients go along with and ignore situations they intuitively know can only lead to failure.

My instinctive reactions to my attorney had become negative early in my experience with him. However, not until the day before the trial did I face the fact, much too late, that I should have fired him long before. Though I was no more capable of making that final decision when I did, the road to failure had become so obvious, I had no choice. The price I would have to pay was too high not to do so. I ended up making the decision to settle the case rather than go to trial to avoid the stress that would follow if I didn't settle.

Shortly after the accident, when he was hired, the prognosis was that I would get well, with no aftereffects. I was told I would be able to return to my job within two months. I was assured the only reason it would take that long was because I had so many physical injuries it would take that long for them to heal. As previously stated, there was never any indication, mention, or consideration about the possibility of the head injury and the possibility of problems developing from the concussion.

Based on this diagnosis, my attorney took care of all the routine legal requirements. He filed the necessary papers and contacted the insurance representatives. But, shortly thereafter, the medical situation changed. Symptoms developed that made it impossible for me to return to work as anticipated. My attorney did not change his attitude, legal approach, or follow-up plan on the basis of these new symptoms, rather he continued to view the case based on the original, incorrect conclusions made by the physicians during the first weeks after the accident. I, of course, did not know this at the time.

Years later, I was to read that a substantial population of patients who appear normal in CAT scans and MRI studies actually have local abnormalities, local injuries to discrete parts of the brain, or severe neurologic disabilities. Counsel should not rule out a possible brain injury simply because a neurologist, a neurosurgeon, or even a neuroradiologist has not made such a diagnosis. Medical specialists use sophisticated diagnostic tools, but they can only identify *macroscopic* [large enough to be observed], as opposed to *microscopic* [too small to be observed] injuries. There may be hope for future victims because lawyers are being alerted to the fact that these injuries, though difficult to detect, do exist.

My medical and physical problems dragged on for almost three years, as did the case. None of my symptoms showed any signs of improving. I could do less and less work at home or at my office. My mind was less and less clear; I was less able to concentrate and less able to retain information. I had become almost incompetent at my office, to the point that my job was in danger, and I could not to help my husband with his manuscripts.

The information the insurance companies and my attorney were using as a basis for their conclusions, taken seven months after the accident, were depositions, testimony under oath in answer to questions by attorneys prior to court proceedings. My physical deterioration for the next two years made no difference in my attorney's view of or approach to the case.

I was convinced that the resultant physical and psychological changes that developed did alter the nature of the case. My attorney did not see it this way, a fact I learned too late. None of this was important to him. Years later, the clue to why he reacted the way he did came as a shock to me. When my injury was finally diagnosed as a brain injury, just before the trial date was set, the response of my attorney was, "A brain injury is just another name for a headache." I now understood why his advice to me all along had been not to see any more physicians, because "It's not good for the case."

With these attitudes, he saw the case as nothing more than a routine automobile accident that had caused no serious injuries, and thus he paid no special attention to the details. His attitude or legal approach to the case was never altered in spite of the new circumstances, and thus there was no preparation on his part.

I can attribute my earlier decision to continue to cooperate with this attorney to the fact that I had forced myself to believe that what he was doing and saying had to be based on knowledge, and thus it was the right thing to do to go along with it. But I wondered if my compliance was the result of not being well or capable enough to make sound decisions. Though my husband and I talked about this, he too, just as with my physicians, believed we had hired a capable attorney. Obviously, all of our judgment can be questioned, in spite of our intelligence.

A few days before the trial date, we still had not had a meeting about procedure. As I write this, it is inconceivable that what happened could have happened, but it did. We had to call the attorney to remind him that the trial was within a few days, and we had never discussed the details of the proceedings. Apologetically, he said the earliest "free time" he had was 4:00 P.M. the day before the trial. I soon learned how foolish I had been to assume he would be prepared to explain the details of what would happen in court, how the procedure would go, what the defendants would do, what I would have to face, who the character witnesses would be, and his general approach. I sat there dumbfounded when I learned he had no character witnesses. Even more important, I had expected to be alerted to the type of inevitable questions that would be asked of me and given some clue as to how to answer them.

All this turned out to be wishful thinking, another step on the road I had allowed myself to follow since we hired this attorney. It was at this point that he told us that another attorney in his office was "working with him" on the case. The true situation soon became obvious. He had turned the entire case over to this second lawyer—a man we would never have hired, under any circumstances—and, as it soon became obvious, even less qualified than the one we did hire. We did not meet him until that day, too late, I thought, to do anything about it. One more client problem: not being knowledgeable enough to know what to do when faced with an unanticipated situation, although logic seemed to require discharging that firm and hiring a new attorney.

When my husband and I arrived at the meeting to discuss whether or not to go to trial, only one attorney was present, the first man we hired. He would not discuss any details about the trial. He was waiting for the other attorney to join us. The one decision I made at this pre-discussion encounter was that I definitely wanted to go to trial, no matter what. I made it abundantly clear to him I was vehemently against settling the case out of court.

When the second attorney finally arrived, harried and short of breath, he managed to find a second to say hello then immediately went into a tirade about a case he had just lost, because it was held in the county court, where "settlements are low," but if it had been held in the city court, he "was sure he would have won." Our case was, of course, being held in the county court. I was not too ill not to recognize what was coming.

A sensible person would have had to be asleep to miss that this was a preamble to a game plan, with two poker players trying to hide the clues to the cards they held. The scene was too polished to believe it had not been rehearsed, ready to be played out to assure them that they would accomplish their objective. They had no plans to go to trial. They were not prepared to do so, and they had to see to it that I didn't want to do so either.

Though my interpretation of this charade was correct, I refused to go along with it. I firmly believe that a trial would prove I was not a malingerer, and this was more important to me than any money I might or might not receive. And so I asked questions that went to the core of the problem, something they did not expect, did not like, and had not prepared for. Perhaps that was why, whenever I asked a question, instead of answering it, my attorney would respond by saying I asked too many questions. He assured me there was nothing to worry about and said he was there to "take care of all the details." Did he think I was not capable of understanding the challenges involved? Or would the answers my questions required have been too revealing of what was really going on?

Questions touch the ego of the person being queried, and answering often requires an explanation that someone is unwilling to give; this makes it necessary for the person being questioned to give the reasons for conclusions which, all too often, the intelligent questioner should not be willing to accept.

Why I expected a different reaction to my inquiries this time I do not know. It soon became obvious there were no acceptable answers to any of my questions. Had they compared the depositions, police, hospital, and medical records for any discrepancies? My instinct had told me there had to be many clues that would have been in my favor. Had they talked to the people with whom I had worked for more than fifteen years to confirm an unblemished attendance record and an excellent reputation for trustworthiness and reliability? Did they have any character witnesses ready to testify on my behalf? They did not. Even in my debilitated state, I knew that with the implication of malingering being present, without character witnesses, there was no case. There was no way they could have been unaware of this.

My questions could not be, and were not, answered by either attorney. The second attorney rummaged through his files, appearing to be searching for information to be able to respond, which, of course, was not there. While he rummaged, my attorney deflected the conversation so the questions never had to be answered; and, indeed, they were not. Neither attorney had properly prepared the case. No one had been contacted, so there were no character witnesses. At this point, I realized that neither attorney had taken the time to familiarize himself with the details of the case.

The inability to present proof that I was an honorable person of high ethics—that my attitude, work habits, and conduct had always been honest, direct, and productive, all my life—would have played into the hands of the defense. Capitalizing on this paucity of medical evidence regarding the residual problems, even the most immature and inexperienced attorney would have been able to portray me as a malingerer interested only in a large settlement. As ill as I was, I was not so ill as to be unaware of this obvious problem.

As I sat through this "no-information" discussion, I could feel my strength waning. I was ready to give up, an attitude I had been fighting against ever since the accident, one that was alien to my basic nature and character. I was not strong enough physically to continue this fiasco. I knew I had to end it, but how? I sat there dumbfounded, as I heard my attorney say, contrary to what I had made so clear to him before his "assistant" arrived: "We had already made the decision to settle the case and not go to trial." This was the final straw, and it led me to my final decision.

I could not believe my ears, although the reason for the statement was obvious. No attempt was even made to hide the fact that he lied; that they had no case prepared, no evidence in my favor, had done nothing. I sat there dumbfounded, a prisoner of those who were supposed to be liberating me. In hind-sight, these attorneys were hired on a contingency basis, so if I lost my case they wouldn't get paid, but if I settled out of court they would receive payment for their time.

As I heard my husband try to counter this outright lie, I stopped him. I told him I had made the decision to settle the case and accept whatever offer the insurance company made, an offer that I eventually learned more than adequately compensated the attorneys but did not come anywhere near covering my medical costs and salary losses, not counting any other expenses. I could not wait to get out of that office, away from these men. I stopped my husband from still trying to respond. He was stunned, but as he told me afterwards, seeing how distraught I was, he was more worried about me than the low settlement. For me, the horror was that by accepting the offer, I would be, in effect, admitting I was a malingerer. If anyone was aware of how distasteful this was to me, it was my husband. But I was insistent, and so I left him no choice but to go along with my decision.

Why did I do it? It had become so obviously foolhardy not to face the fact that I could not undo the damage these men had done; that going to trial would have been an utter disaster, not only for the case but, more important, for me physically. Continuing to fool myself into overlooking what I had been aware of for so long was no longer possible. The alternative would have been to fire them then and start all over with another attorney. But I could not afford to waste what little strength I still had and thus could not consider that alternative, giving up seemed to be the only reasonable choice available to me at the time.

As I said, the settlement never covered my financial losses, but the fees these attorneys collected more than compensated them for the time spent and services rendered—or, more accurately, not rendered. I lost my job and spent the next few years trying to get well, but I did not succeed. Through the years, I never gave up searching for answers to how such things can happen, how attorneys can get away with such incompetence, and where the responsibility lies for such legal representatives being allowed to practice law.

PART TWO: The Lawyers' Dilemma

Was I merely a dissatisfied client, viewing circumstances and actions from my personal bias? Was I unhappy merely because the settlement was too low? Or was I correct in my interpretation of past actions and inactions and what I had seen at that last meeting?

Why are there misunderstandings between clients and attorneys? Are legal conclusions purely personal ones and intellectually logical ones based on facts? What causes these problems, and why are they so prevalent? Are clients expecting too much, or are attorneys doing too little?

The legal profession, well aware of its responsibilities to its clients, is equally concerned about the damage to the profession's reputation when attorneys do not live up to their legal responsibilities. Whether the failures of some attorneys are due to incompetence, special circumstances, misunderstandings between client and attorney, or the myriad problems that abound in all complicated situations, it does not alter the need to try to limit these failures or to attempt to develop some means of recourse for the injured client.

Many in the legal profession have recognized the problem and proposed solutions. In 1954, Chief Justice Earl Warren charged that many trial lawyers are unskilled and incompetent and, despite law school tests and bar examinations, said that "we are more casual about qualifying the people we allow to act as advocates in the courtrooms than we are about licensing electricians." He added, "Imperative and long overdue is some system of certification for trial lawyers."

Then president of the American Bar Association (ABA), Chester Smith, pushed for requirements that lawyers periodically and regularly demonstrate their competence in order to continue practicing law. Though standards of competence are a rising concern in the profession, it is the question of ethics that touch the most responsive nerve among lawyers. Could this be because incompetence is a reflection on all attorneys, but morality is not? In 1973, the ABA's Center for Professional Discipline was established. Its Committee Chairman, S. Shepherd Tate, stated: "The Center receives substantial numbers of inquiries from persons who have had real problems with lawyers and is benefiting from experience gained in these contacts. Bar leaders are fighting hard for discipline changes needed to retain public confidence in lawyers."

"The understanding is growing," said Tate, "that public confidence in lawyers is important. Bar leaders are fighting hard for discipline changes needed to seek this confidence. The license to practice law is like most others; its abuse must cost the offender some or all of his rights under the license." The alternative might be outside interference, something no one wants.

"The public perception of lawyer discipline is at odds with the more benevolent view of a majority of lawyers that complaints against lawyers are frivolous and the penalties draconian." (Peter Brown, 1992). The same article included a comment about Robert McKay, a spirited law teacher who, in 1990, alerted the legal community to the fact that our disciplinary system requires "substantial reform," and that among other disciplinary violations was listed "greater client neglect." One problem, of course, is that disciplinary systems, which vary in performance, are generally underfunded, overloaded, and deprived of expert staff.

In 1995, a local bar association in a Midwestern city asked approximately 2,000 of its member lawyers what they thought the public thinks of them. The answer was "Not much." Sixty-three percent said they had seen incompetent representation; 38 percent said they had seen other lawyers engage in unethical behavior. Only 28 percent said serving the client is one of their most important goals.

The conclusion, according to the president of that association, was that the survey provided evidence that the public's mostly negative perception of lawyers was not based on myth or misrepresentation. Though the professional code of lawyers requires them to report improper or unethical behavior by their colleagues, only one in eight did. Unfortunately, this same response can also be applied to the medical profession. Another somewhat surprising statistic was that three out of four attorneys gave themselves only a "fair" rating on twelve aspects of their work.

These statements were made as early as 1954 and as late as 1995, evidence that the problem does not lend itself to ready solutions that work. Though both the legal profession and the public seem to believe there are more competent attorneys than not, appreciating the dilemmas that exist in attorney–client relations makes finding solutions imperative.

As far back as 1965, Richard Sims III, a San Francisco attorney, declared: "I feel lawyers are a hell of a lot more honest than the vast majority of the clients they represent." If he is right, and the client is doing something unethical, what is the attorney's responsibility? Should he take the case anyway? If he does, would this not result in a negative and detached attitude by the attorney toward his client? Does this not encourage an attitude that makes viewing the client as the problem rather than the problem being the reason for which the client has hired an attorney? If he knows the client is doing something illegal or unethical, how does he morally and intellectually cope with that? Does not handling enough of such cases alter one's own view of right and wrong?

"We try a lot of cases by the seat of our pants, interviewing the client as we go into the courtroom," an attorney once observed. When the important legwork is not done prior to trial, what alternative is there but to convince the victim to accept the insurance company's offer, rather than go to trial—a form of an assembly-line mentality? With this attitude, it is not difficult for the attorney to develop an emotionally detached method of dealing with the client and the details of the case; stereotyping his client, minimizing his own involvement, and relying on formulaic solutions.

Not until after the case was settled did I have the proof of how little was done. There were discrepancies in the police report, the hospital records, and the physicians' letters. Nothing had been checked or verified. There was no indication that any effort had been made to confirm who was actually responsible for the accident. The result encouraged each insurance company to blame the other driver. It was obvious why neither one would accept responsibility for agreeing to a fair settlement. If my lawyer had done anything to dispel this misconception, there was no evidence of it.

Rarely are there simple solutions to legal problems. Explanations must be difficult for an attorney to express clearly and equally difficult for the client to understand or accept. I never had to worry about this.

At no point did my attorney ever explain the complicated problems of the case, nor was I ever told why I should or should not be prepared to go to trial. Choices were never delineated, and so it had never occurred to me that we would *not* go to trial. But my attorney could not have been less prepared to try the case than he was. If I in my debilitated state could see it, surely it would have been clearly visible to the defendant's attorney, had they not already become aware of it from the discussions that must have taken place prior to the anticipated trial date.

The drivers' depositions covered their roles to throw blame on the other driver. I guess this is normal procedure for the defendants, but what was normal procedure for my attorney? Shouldn't these details have been verified? For example, one driver said he had not been drinking. But had he? He was a salesman who had just taken a customer out to dinner. Verifying whether he had or had not been intoxicated at the time would have made a major difference in the final conclusion about responsibility, since he was driving the car that hit the car I was in.

Even with all the possible relevant information, my attorney would have had difficulty representing me fairly at any trial against the insurance companies' attorneys, who had everything in their favor. With no pertinent information to counterbalance theirs, the result was inevitable. When this became obvious to me, I was left with no choice but to make the decision I did.

An attorney who is unable or unwilling to mount an investigation, research applicable law, explain options to the client, and prepare for trial does not live up to his responsibilities; he isolates himself from his client and ignores important facts. Believing his approach is valid, based on his past experience, his legal responsibilities can readily be sacrificed.

As a client, ill or not, I probably did not live up to my responsibilities. I should have paid more attention to details, asked more questions, and insisted on understandable answers. I often felt as though I was on one side of a sound barrier with my lawyer on the other; I felt absolutely deserted by him.

Getting good legal advice is a combination of luck and knowledge. The wisdom to know when you are not getting it and having the courage to do something about it when you are in doubt is not easy to come by. How and why my case turned out as it did was and is still appalling to me. Was I the greedy client? Was my attorney competent? Or was it the other way around? Losing or winning a case depends on luck, power, personalities, and sometimes even on the pressures of the court calendar.

Winning a case is important, just as a fair settlement is. When that does not happen, it is distressing. But when the knowledge that winning a fair settlement didn't happen because of the incompetence of the attorney is added to that distress, the reaction of the client is often anger, especially when the law states that an attorney's compensation must be paid even though he has not earned it.

PART III

Bar Association's Dilemma

CHAPTER 13

The Insurance Involvement

Insurance: coverage by a contract to indemnify another
against specific loss in return for premiums paid.
American Heritage College Dictionary 1993

It is better to be roughly right, than to be precisely wrong.
John Maynard Keynes (economist)

Insurance is thousands of years old and has grown like topsy-turvy. In
1700 BC, the Code of Hammurabi in Babylonia included a form of credit
insurance. In ancient Greece and Rome, burial, pension and disability
insurance were available. During the Middle Ages, guilds added fire and
theft insurance to the list of available services. In 1690 marine insurance
and underwriting began in a London coffee house owned by Edward
Lloyd... forerunner of today's Lloyd's of London.

The theory of probability, still widely used in determining insurance
rates, was developed by two French mathematicians, Blaise Pascal and
Pierre de Fermat. The English astronomer Edward Halley developed the
first mortality table in 1693, and Benjamin Franklin founded the first
mutual fire insurance company, *The Philadelphia Contributionship for the
Insurance of Houses from Loss by Fire.* In 1759, the Presbyterian Ministers
Fund became the first life insurance company in the United States. Both
companies are still in existence.

In 1864 the Travelers Insurance Company, the first accident
insurance company, covered a Hartford, Conn. resident, during a two-
block walk from his home to the post office. The premium was 2 cents
for the walk. Could this be where the phrase "2-cents worth" originated?

In the mid-1800s, many states established insurance departments
and passed laws "regulating the industry as a result of dishonesty by some
companies. However, insurance laws frequently were not strictly
enforced. During the late 1890s, the industry was plagued by scandals

caused by the dishonest and irresponsible practices of many companies." (World Book Encyclopedia, 1996). In the early 1900s, many states passed laws that regulated the activities of insurance companies more strictly.

Today, as in the past, insurance requires an alliance between two factions with opposite motivations. The insured purchases insurance for peace of mind but always with the hope there will be no need to collect payment and thus is "unhappy" to have to purchase it. Their "happiness" is dependent on receiving payment when a loss occurs.

The insurer sells insurance for a profit but always with the hope that there will be no need to make payment and thus is "happy" to sell it. Their "unhappiness" is when they have to pay for that loss because even the best insurance companies are businesses and they profit from not paying out claims and when they pay, they hope to settle claims as cheaply as possible. An excellent example of this dichotomy is when a claim for insurance is filed after an injury caused by a concussion...or a hidden injury. This hidden injury is concealed from those on the outside, but the possessor is acutely aware of it. How insurers view this injury and why they view it the way they do are the crux of the problem.

There is an unasked question that needs answering: from whom is it not hidden? It is not hidden from the victim because the pain is constant. Only those people who know the victim well are aware of the pain and problems this injury has caused, though the physical proof is hidden from them too.

"There is only one pain that is easy to bear," said the French surgeon, Rene Leriche, "and that is the pain of others." (U.S. Dept. of Health, Education, and Welfare, Sept. 1968). For obvious reasons, and individually perceived motivation, the patients' descriptions of their pain and symptoms, are interpreted from different professional, cultural, and financial points of view. Pain and related symptoms are so subjective, they defy universal definitions. They are difficult to define for others because they are perceived introspectively by the person feeling them. These messages need to be decoded, interpreted, and acted upon by physicians, attorneys, and insurance companies.

Leriche believed that pain is a disease in itself and is dehumanizing. As a warning, pain by the time it appears, is often too late. It is not perceived until after the disease is well advanced. This certainly applies to the slow, methodical development of symptoms after a concussion.

Courville wrote, in 1953, the concept of commotio cerebri (concussion) can be traced back to the 16th and 17th centuries, but was of little concern to the medical writers of the time. A concussion of the brain, as late as the 1860's, was viewed as a "certain indefinable something, a cause of evil which could not be demonstrated." Words like Commotio, succesio coerebri, or enbranlement applied to the phenomenon of concept of movement, or even double movement (coup contre-coup) first voiced in 1767, hiding the actual nature of the injury in these difficult to comprehend terms.

Is our use of substitute terms like post-concussion syndrome, post-traumatic syndrome, MHI, functional, for concussion or brain injury a recapitulation of what occurred in the 1800s? Whether motivation was the same is open to question. The results pertaining to the knowledge conveyed to the patient are not.

More than one school of thought exists on this subject today, but the known facts upon which there is agreement are: there was a head injury; there was a concussion; the person may or may not have been unconscious; the victim has no memory of how the injury occurred. Thus it has become necessary in recent years to relearn what medieval physicians suspect caused them.

With the rapid development of industrialization, railway and automobile collisions increased. People were exposed to the high-speed acceleration-deceleration injuries (coup contre-coup) resulting from these collisions. Thus, in the 19th century, the study of concussions began. The unequivocal conclusions were that concussion was a genuine entity, a physiological disturbance. The more recent research in the 20th century confirms this conclusion.

The litigation that developed from these troublesome injuries caused medico-legal problems that probably helped contribute to the greater interest in this injury. This litigation inevitably led to the "logical demand" for receiving compensation resulting from the accidents that caused these injuries...compensation based on the degree and extent of the injuries. When a fractured skull or cranial trauma injury is obvious, and neurological or psychiatric deficits develop, there is no ignoring them. Proof of the "concussion" cannot be denied. Not so with "commotio cerebri." It is difficult to confirm the symptoms.

"The neurological examination is often essentially negative, the ordinary roentgenogram (x-ray) of the skull is valueless, and even the electroencephalogram often fails to confirm the occurrence of physical injury to the nervous system. Because purely subjective complaints are difficult to evaluate, the insurance agencies have not been slow to seize upon this fact and argue that because nothing can be demonstrated objectively it should be assumed that no disability has actually taken place. Because of this, a school of thought has arisen in both the medical and legal professions which implies, if it does not specifically so state, that the symptoms so often complained of consequent to an episode of concussion constitute nothing more than a psychogenic disorder inherent in the patient's deficient personality (traumatic neurosis) if he is not actually guilty of an effort to secure gain by fraud (malingering.)" (C. B. Courville 1953)

Setting aside financial motivations, what medical dilemmas added to the disagreements of the true nature of this injury?

Jean Martin Charcot (1825-1893) director of the famous Salpetreiere Hospital in France, was widely regarded as the principal founder of neurology. He investigated many aspects of the relationship of organic conditions to both physical and mental disorders. At that time, hysteria was exclusively associated with females, primarily because the Greek root hyster refers to the uterus, another myth regarded as fact. Charcot, however, noted that some males suffered from hysteria, and thus the problem could not be explained in terms of female anatomy. (Frank J. Bruno, Ph.D.) This is a fact often not readily acknowledged even today.

However, Charcot (1889), a profound student of the problem, confused the post-concussion syndrome with those of neuroses. While many in the 1890's were beginning to recognize a distinction between true post-concussion syndrome and functional disorders, others failed to differentiate between the two, a problem still plaguing us today.

How much responsibility can be placed on the shoulders of the insurance companies when, in 1974, they assigned an orthopedist to validate - or invalidate - my claim, and he failed to differentiate between the two? His conclusion was expressed as "in my opinion." He was not only wrong in his opinion but he had no data to confirm that statement, which was based on the inaccurate diagnoses of my previous physicians.

Unless there is a specific motive, why would an insurance company assign an orthopedist to evaluate one's inability to function at one's prior level of capability? I knew, they knew, the physicians knew my bone injuries had healed at least a year ago. But even more important, is an orthopedist qualified to evaluate whether the "head" problem prevented me from functioning normally? Would the insurance company have given this physician my previous records if these conclusions stated there was something seriously wrong with me? Would this orthopedist have automatically confirmed those conclusions or would he have searched for reasons to invalidate them? Is there any doubt about the answer to this accusative question?

Because, many manifestations of psychic character in patients with post-concussion syndrome are also found in neuroses, the two can be and often are easily confused. This confusion has resulted in the assumption of many that all post-concussion manifestations are purely psychic in origin, predicated in most instances by conscious or subconscious desire for gain. Since the motivation of insurance companies is to settle claims for the lowest figure, it is not difficult to understand their attributing the same attitudes to insured individuals assuming trying to get the highest figure has to be the only motivating factor.

It seems irrelevant to them that the symptoms labeled "post-concussion syndrome" occur universally, in patients who are unknown to each other; that the same reactions occur in patients who have no hope of compensation as they do in those who do; that these symptoms occur in children who do not understand the intricacies of insurance; that they persist in injured individuals even after entirely satisfactory settlements have been made.

When insurance companies as well as some physicians and attorneys question the victim's motives, their views have led to serious and often tragic miscarriages of justice for individuals that have sustained injuries to the head. Could this be the reason why, on the other side of the coin, juries, aware of this avoidable injustice, have, in rare but publicized cases, granted awards unfair to insurance companies?

Although the two neurologists were responsible for laying the groundwork for others to believe this subtle accusation against me, it is important not to ignore the other side of the question. The presence of many obvious and often prominent manifestations of physical characteristics in patients with post-concussion syndrome is responsible for a misunderstanding of the true nature of this state. Many of these symptoms are also found in those who do have a neuroses and is the reason why the two conditions are easily confused, particularly by the untrained observer.

There is a wide range of different degrees of knowledge about concussion and prejudice for or against the concept that its residual symptom-complex constitutes a true clinical entity. Courville's hope is that the evidence has been such as to convince the most recalcitrant of his readers that concussion can produce disorders of the brain which are both acute (due to the shock of the injury) and chronic (due to the circulatory system), and that true post-concussion manifestations sometimes are simulated by post-traumatic neuroses, either independent of, or associated with, post-concussion syndrome.

Two schools of thought continue to persist, but rational observers should be able to have a clear concept of the relationship between the two.

> *"What was accepted as fact a half century ago has become (is) a moot question in the minds of industrial accident commissions and courts of law (a half century later.) Many an individual has been (is still being) denied any redress for his injuries on the grounds that no objective evidence of his symptoms could be produced." --Courville 1953.*

Procedures are being, and must continue to be, devised by the caregivers to find practical, clinical methods to prove or disprove the presence of concussion and its disturbing residuals that result in the severity and duration of symptoms. The legal and insurance professions must open their minds to the importance of "correct evaluations rather than beliefs and subjective conclusions." These are vitally important elements for the well being of the injured, the professions, and society if justice is to be administered.

The physician who believes the patient is neurotic will be 'indifferent' in the effort to make any serious attempt at therapy. If he believes the patient is a malingerer, his sympathies will be even more tenuous. It is therefore necessary to confirm whether the patient is suffering from genuine symptoms. If no question of compensation is involved, the physician is not so likely to be prejudiced against the victim. When compensation is involved one should be alert to the possibility that prejudice may be present, in which case a more critical attitude should be assumed, and perhaps, second and third opinions solicited from unbiased physicians.

But in all fairness, just as it must be said that the victim should certainly be protected against those who would seize such an opportunity to pay as little financial compensation as the traffic will bear, so too it must be said that the insurance carrier should certainly be protected against that group of individuals who would seize such an opportunity to secure as much financial gain as the traffic will bear. Being fair is not a one-sided responsibility.

How much do circumstances have to do with how the injury is viewed, diagnosed, and treated? With no clearly delineated "medical standards," or "guidelines," individual beliefs, personal prejudices, unconfirmed conclusions are stated and often accepted, depending on the predilection of the recipient of this information.

> *"The assumption underlying dissemination of guidelines, as well as continuing medical education, and the publication of scientific journals, is that of "rational choice": that physicians change their practices in response to new*

information. In reality, it seldom works that way. Physicians are slow to change their behavior until they perceive a need to change, and this perception of need is more likely to occur in response to social than informational influences." --Lucian Leape, M.D., JAMA, 1995, Page 1534.

The complaint on the part of insurance carriers or their employed physicians regarding the frequent occurrence of post-concussion neuroses in complicating the picture of recovery, have largely themselves to blame. Courville states he has had repeated opportunity to verify this conclusion. "The patient's complaints of quick brush-offs at the hand of insurance physicians who insist that they can do no more in return for what they are paid by insurance companies are quite typical." This statement, written in 1953, answers my questions about my insurance doctor.

If I, as a patient, had asked a physician to write a report in my favor, I would be hauled into court as dishonest, fined or jailed. If I am not permitted to do this, why should the insurance company and/or the medical profession be allowed to do so, even though, of course, they cannot admit that is what they do?

It is as if insurance companies are allowed to practice a form of medicine without a license. Today, this problem is rearing its head more obviously in the HMO/physician/patient present-day controversy, with the government concluding we need a "Patients' Bill of Rights" to assure fair and proper medical care through HMO's.

When multiple errors occur in seemingly disconnected circumstances, there are similar connections.

I was unaware of the seemingly "disconnected" circumstances in my case until the evidence made the connection obvious and impossible to ignore. It was not by accident that the first neurologist's attitude struck a familiar cord with the second neurologist, or that both their conclusions struck the same cord with the orthopedist, who were both in the employ of insurance companies. The basis for the orthopedist's "medical" conclusion that the problem was "functional," was not based on his medical evaluation but on the reports of the other doctors who had come to the same conclusion. The fact that their medical conclusions had no confirming "standard operating medically required information" and might have been inaccurate did not seem important to him.

When defendants are told they must see the physician who represents the insurance company, they are often unaware, or have given it no thought, that this physician has all their medical records with the diagnoses of the previous physicians. Loyalty to his employer "requires" motivation to take precedence over proper medical procedures. This is but one additional step in the iatrogenic result of a wrong diagnosis.

His final report to his employer, the insurance company, did not even mention that there had been a concussion. This omission was covered by the statement, "She was not thrown from the car but she *believes* she was unconscious."

Again, the question of conflict of interest comes up. It seems so obvious in these episodes yet these actions seem to be accepted medical practice. Questioning this orthopedist's conclusion led me to again follow up for answers. A contributing editor to JAMA responded[*] and concluded his letter with "I apologize in that these answers are somewhat general, they are probably about all you will be getting. I hope they will be helpful."

What was interesting about this response was that although it was written on AMA letterhead, with JAMA indicated below, when I questioned whether this was the opinion of the AMA I was advised, "The basis of my response to you...was as a concerned physician acting as a patient advocate for the 40 years I have been in practice. My opinions were personal and represent only my own conclusions."

Dr. Courville was well aware of this and other medical and insurance problems in 1953. In 1972 we were no closer to correcting these actions, and it seems we still have a long way to go to do so.

"The extreme "neurosis school" is kept alive by some physicians, employed by insurance companies, who refuse to see any but their own side of the question. A critical examination of the individual symptoms and the patient's background, though this is admittedly subjective evidence, are essential factors required to distinguish between the true concussion and the "malingerer." How can this undesirable state of affairs be remedied?

The accumulated evidence of experiments and investigations that the residual symptoms are due to disturbances of functions of various parts of the brain is such as to form a solid basis for the thesis of the reality of concussion. In the aggregate, many "less obvious signs" may be important but are equally dismissed as insignificant. There are a host of "minor" inclusive findings of a general and neurological nature, that may not be significant as independent observations, but when viewed as part of a group of symptoms, together with the subjective ones described by the patient, should be presumed to be relevant, rather than merely labeled post-concussion syndrome.

In the universe, there is a "black hole" from which nothing that is drawn into ever emerges. Is there a "black hole" in medicine, in law, and in insurance where information about concussions is sucked into and also gets lost?

Among the major reasons why this injury is viewed the way it is are: motivation; disregarding subtle "clues", subtly evaluating the patient or victim to be a malingerer, thus concluding the symptoms from this injury to be functional, and costs. There may also be less understanding of the brain by the professionals than is generally acknowledged by them and the general public.

Facts have tended to confirm some of the earlier observations on concussion, symptoms produced by the abnormality of the function of the brain. Though recovery takes place in a large number of those afflicted, the balance will continue to have symptoms and persistent disabilities whether compensation enters the picture or not.

In an indirect fashion and through continued research by those who refuse to hold on to preconceived beliefs, there is support for the conclusion that concussion in the twentieth century is just as genuine a clinical state as was commotio in the sixteenth when the concept of compensation was unknown.

While purely psychic, post-concussion manifestations may occur, the counterfeit often is confused with the genuine. An intelligent, comprehensive and up-to-date therapeutic approach to the problem must be found to help accurately differentiate one from the other, to help bring relief of patients' symptoms and hopefully lower costs to insurance carriers. If one searches diligently it will find, even if not a total solution, at least enough data to require rethinking present beliefs.

There is a blindly accepted scenario in personal injury cases that is acted out as though someone had written a flexible script for it. The participants in this "silent conspiracy," (or alliance, or holy alliance, or unholy alliance) subconsciously make minor adjustments in their thoughts to fit the individual circumstances of each case. Accepting the flexible script makes this possible. Using statistics to bolster their conclusions, they believe that 80 percent automatically get well and so assume the other 20 percent are responsible for not getting well. However, The Mayo Clinic Family Health book statistics include a statement that "about a third have a continuation of symptoms."

Two concepts contradict each other; 1. The knowledge that the victim did have a concussion. 2. That there is no brain injury...in spite of known symptoms that developed after the concussion. The first requires medical care, time and money. The second eliminates those needs and helps sustain the belief that the victim is either a neurotic or a malingerer interested primarily in a higher settlement, or both.

Barry Willer, Ph.D., in an editorial in i.e. Magazine, Vol. 3, Issue 4, 1996, wrote: "At a public hearing, I heard an administrator of an important health care agency, state: '...there is no such thing as a mild brain injury.' His belief, like so many in the insurance, as well as medical and legal fields, is that the problems witnessed after a brief period of unconsciousness are accounted for by malingering.

"Brain injury renders most survivors incapable of realistically assessing their overall capacity to function. Practically all have clear memory of themselves as they conducted their lives premorbidly and a self-concept based on this remembered functional capacity. All require a period of clinical intervention that guides them towards an awareness of their residual deficits and strengths, helps them acknowledge the limitations imposed by these deficits, and encourages them to make accommodations -- in the form of compensatory measures -- for these lasting incapacities. The period of prevocational preparation is designed to close the gap between how they were pre-injury and how they are now.

"The content of this preparation period may vary in style of presentation and manner of implementation depending on the level of post-trauma functional (work related) skills as well as the patient's capacity for understanding or conceptualizing the impact of deficits on occupational, academic, and social reintegration." (Silver and Kay, 1988).

Not accepting this view encourages the adversaries to be unconcerned about their roles but still viewed as upright citizens with no ulterior motives, and does not hold them responsible for ignoring relevant medical data.

What is the role of insurance companies regarding "fair settlements" or "low settlements" regardless of the merits of the case?

Max Wildman, an attorney with a prestigious law firm whose clients include many of the biggest U.S. insurance companies and corporations, in a personal interview, proudly stated he appears in court shabbily dressed, one of the many ploys he uses to get the sympathy of the jury. "Why should the other side have a monopoly on sympathy?" he explained.

Mr. Wildman handles only a half-dozen of the biggest cases, out of approximately 1600 that the firm handles. "He gets the kind of cases where the plaintiffs are so horribly injured and the liability so clear that the insurance company has little chance of winning," explains a fellow attorney. "They want Max because they know that there are few defense lawyers better at holding down the size of the jury verdict...And sometimes Max even wins the case."

Mr. Wildman explained his approach to jury selection. He selects elderly "retired people living on fixed income and older blue-collar and middle management workers. They are accustomed to shifting for themselves and are usually conservative with awards." He also frequently resorts to artifices and ploy. "You can put on the strongest case, but if you don't use ingenuity you'll get murdered by the jury anyway," was his casual, unapologetic explanation for this attitude. "I can't tell you how many cases I've won exploiting confusing meanings," his method of applying his knowledge of semantics. (Wall Street Journal, For The Defense, Jonathan R. Laing 1973).

With a slight stretch of the imagination and this example to confirm the oft-stated attitude of insurance companies of their inability to differentiate between "real and fraudulent claims," one wonders if they have the same difficulty and concern differentiating between "real and fraudulent presentations" in court when their own attorneys are involved? With the success of this and similar scenarios, instead of receiving a fair settlement, the victim is left to cope with financial worries that create psychological, physical and emotional problems that might have been alleviated, if not avoided, had they not had to cope with the additional burden of convincing the professionals that their symptoms are real.

The predisposition of the injured individual may be anticipated by adequate, yet simple, investigation, and proper preventive measures before the seeds of injury are sown and watered by emotions and apprehensions and fed by professional indifference to perhaps eventually mature into fruitful neurosis, as enumerated by Courville.

It is my hope this book will serve as a stimulus for a more critical study and acceptance of the authentic character of the hidden injury and its role in leading to the post-concussion syndrome. It is also my hope that the members in the professions involved in this care will include in their study the awareness of the need for a more clearly understood terminology to eliminate the confusion about the injury and the incapacitating results that develop from it. This would be a most impressive demonstration to those members of the Functional School that post-concussion syndrome is deserving of a clearer terminology than is this unclear, misunderstood and misapplied description of a serious problem.

The hidden costs for this hidden injury...medical, financial, and psychological...for the victim and for society are prohibitive. Cooperation and fair treatment for all can only help eliminate the unfairness of the present system, and in the end, lower the costs to society.

PART IV

A Look to the Future

CHAPTER 14

A New Life

You are as you perceive yourself not as you are.

–Author Unknown

The path from despair to hope had always seemed endless, to success, insurmountable. After years of living with this thought, and with my changed personality, and after being unable to do anything about it, I awoke one day to find myself refusing to continue to live that way any longer. Before I could proceed on what I knew would be a difficult experience, I believed it was essential to evaluate my life to try to understand how I had arrived at where I found myself. I needed answers to all those questions that had been present since the time of the accident that I had not yet faced.

Had I accurately assessed the person I thought I had been before the accident? Was I correct in what I thought I had become afterward? Who was I now? What had caused me to go from being a proactive personality to a passive one? Had my sense of reality been distorted? If so - why? How?

My view of reality before the accident had been based on rational thinking, knowledge, and the ability to understand situations and find solutions to whatever challenge I confronted. I never doubted my view of reality, so deciding on my values and goals was a simple task. However, my view of reality after the accident was based on a misconception that there was no medical reason for the problems I was having. Regulating my behavior accordingly became a difficult, if not impossible, task.

To identify beneficial or harmful aspects of reality, one must be able to evaluate them. How I evaluated my situation was the link between reality and my actions. But any reality based on false premises can only result in wrong conclusions, which I had in great abundance. But the question that had to be answered was, who am I now?

When they work in harmony, circumstances, time, and age can effect amazing changes, positive or negative, no matter how difficult the path may be. I was now the unhappy possessor of a dual personality: I recognized a proactive, pre-injured me and a passive, post-injured me. The normal, competent person capable of feeling deep and healthy emotional responses was gone; that part of me was unavailable, just when I needed it to help me cope with my new situation. I could no longer handle even routine problems. While my husband was still alive, he would help me. He had been the center of my life for almost 50 years and had always been there when I needed him. After his death 20 years after my injury, all that changed; it left a deep, hollow feeling and no choice but to find a way to face the future. There was no place for negative thoughts. I had to learn to concentrate my energy on replacing such thoughts with positive ones. Changing my habits, my thinking, my attitudes, my hopes, and my needs became essential, not an easy task after so many years of not being able to do so under less difficult circumstances, even with my husband's help.

My meandering thoughts took me back to memories of Max and his profound wisdom, and I recalled the pleasure of his company. I remembered his personality and how others viewed him; I thought of how we worked together on his books, each evening after a full day at our offices; of the people we met because of the popularity and success these books received; of how he appeared on the lecture platform and the responses of his audiences; of his knowledgeable and inquisitive mind and what he did with it; of our family and what we meant to each other and what he meant to me; and how much I appreciated the role he played in my life. Whatever strength I have now is due in large part to him.

That relationship planted a solid foundation for the future that became even more important, even more indispensable, as events unfolded. Sadly, sight of the obvious often does not become apparent until long after the point at which we should have been cognizant of their importance. These thoughts helped me develop an awareness of the importance of memories, and I had a strange premonition that the solution to my problem lay in my past. Nestled in my mind was an uncanny feeling that paying attention to the characteristics that had been responsible for my success in the past would be helpful. Accepting this thought was the first step that led me to where I am today, a thought that gave me the strength to start on the unknown road ahead of me.

Even in his absence, my husband came to my rescue. I was reminded of how often he would reprimand me for "underestimating my capabilities."

Today, he would be proud of my accomplishments, but he would not be surprised. I was surprised when I found I could accomplish what I never believed I could.

Important decisions had to be made, and soon. Something productive had to be found to fill my time. I could not allow the emptiness left by his death to be filled with the inevitable miserable and lonely feelings that would surely follow. We had always spent so much time together, separated only on rare occasions. A time consuming and productive project had to be found to fill that void lest misery and mourning step in, which they certainly would if I did nothing to prevent it.

Depression was one of the secondary effects of my injury. I never did lose the fear of it taking over my life, and that fear was now exacerbated; it made the luxury of feeling sorry for myself, an emotion that often develops when one is depressed, unacceptable.

Instead of facing the truth, I used a form of self-psychoanalysis by "kidding" myself into thinking my husband was on another lecture trip, and I acted as though he were. I kept myself busy as I had always done when he was away, and it helped. I was well aware that I was substituting fantasy for reality, but by procrastinating, I was able to put off any concern about this false picture until I could deal with it later. I told myself it was only temporary, reshuffled my feelings, my thoughts, and my needs knowing full well someday I would have to switch this "wish fulfillment" into genuine reality. Procrastinating made it easier to do so later. Any other action was too painful. Instead of following the popular advice, that one must mourn, I went back to my memories. I was afraid if I did mourn that I would never get over my loss and my depression would be in full control. No one thought I would succeed. They were wrong.

Recapturing those memories of my earlier years with Max carried me through this most difficult period and helped me build a new life without my husband. So, I began my new life, and I started it with my memories. Instead of misery I decided to live with those happy thoughts.

My memories started with the experience that was to change my life: that first meeting with Max, on my first and only blind date. We had four additional dates, within four days, before he was scheduled to leave the states to take up his assignment in Europe during World War II. Since I would have said yes had he had asked me to marry him before he left, my excitement was unbounded when, in his first letter, he did just that.

After a year's absence, Max returned and we were married, the start of 48 wonderful, memorable years together. His job had been in U.S. military intelligence, interviewing German prisoners of war. He was asked to return to continue this work of weeding out German Nazis from the new Germany. He rejected that offer when he learned my three-year old daughter had seen films of bodies in concentration camps and had difficulty sleeping because of it. He was not willing to take her to Germany, fearing she might develop serious psychological problems. Instead, he went back to work for the company he had been with before enlisting, at a salary that meant living from paycheck to paycheck in a strange city, a small price to pay for his new daughter's peace of mind. We had more than 25 years of a happy, productive, and successful marriage before the accident, and I can always be grateful for that.

One unhappy memory that was especially difficult to erase did arise: I wondered how he was able to live with me after the accident. I had lost my ability to love, to laugh, and to enjoy living. I was no longer the person with whom he had fallen in love. I have never been able to understand how he overcame that burden. My memory, which revealed how much less I had become; my inability to respond to his warmth, affection, and concern; the pain I felt then, and the pain I still feel—all minor things compared to the pain he lived with during those endlessly difficult last 20 years of his life. But I was incapable of changing, much as I wanted to. I was incapable of doing anything about my symptoms then, and I am incapable of doing anything about them now. They made me sad then, and they still make me sad today. No matter how hard I try, my ability to recoup normal feelings and responses is gone, never to return. I must accept that, and today, I am the only one who has to live with my cold and indifferent responses. No one else is paying the price my husband did. I am ever thankful for the help my husband gave me, his dedication, persistence, courage, and the positive attitude he also instilled in me. It was what made it possible for me to keep going then, and it still helps me today.

To start doing what needed to be done, I resorted to self-psychoanalysis, a not-so-scientific method, but it helped. I told myself to think about returning to my typewriter, and a voice within reminded me that writing had served as a release from stress before. Was my subconscious telling me I could do this again, just as I had done it after the accident? I hoped so, but I worried it might not tell me what to type. Then, what to type became obvious. After my husband's death, the answer was not so obvious.

The publishers of *Jews, God, and History* had wanted my husband to revise it. It was Max's first book, published in 1963 and considered by many to be a classic, although he never did. Instead of revising it, however, he wrote four other books. The thought startled me as it leapt into my mind and kept telling *me* to consider revising the book. How could I even consider such a thing? I had never written anything myself, although I had worked closely with my husband on all his books. I toyed with this idea, perhaps because the thought was so challenging and gave me so much pleasure. I encouraged the thought while ignoring the important fact that the publishers might not even consider my revision.

The year I worked on the revision accomplished what it was intended to do: it consumed all my energy, filled my thoughts, and left me no time to dwell on my loss. Much to my surprise, the publishers accepted my revisions, and it was published in 1994 (it is now in its tenth printing). This successful endeavor enabled me to look into the future with hope instead of despair.

Free time now allowed me the luxury of finding a competent physician, one who would not view my depression as being caused by a psychological response to the accident and who would know what to do about it. I was lucky: I found one. With his medical training and his acceptance of alternative—or, for a better word, *complementary*—medicine, his treatment led to a change in the degree of my depression. The result was a slow, steady improvement that made my depression less destructive and less of a handicap, even if it is not entirely gone.

His approach seemed to be to apply the Gordian Knot theory of getting to the core of the problem while cutting through the debris surrounding it. He saw my depression as a response to the fact that, for so many years, I had been unsuccessful in reversing the negative effects of the symptoms. With another burden lifted, I was now ready for the next challenge in my search for something resembling happiness and a good life.

It was encouraging to be able to think thoughts that had once been unthinkable, and such a short while ago. I felt capable again, albeit with limitations. These positive experiences and the seeds my husband had unwittingly planted, generated new ideas and helped me alter my view of myself. This enabled me to seriously consider the idea of doing something I had always wanted to do: try to find answers to why an accident, viewed as a minor injury, could have resulted in so much damage to my life, to me, and to my family. An even more important question that had to be answered was how I could have allowed this to happen. How much of it was I responsible for? I had put off answering these disturbing questions for many years, but I could no longer do so. Was my experience unusual? Were there others, not as fortunate as I was, whose husbands deserted them because they believed the injury was not the cause for the symptoms, and their wives were no longer the same person they had married? How many others did not find the physician who would eventually make an accurate diagnosis? I turned to the task of finding answers with verve and energy, uncertain and perhaps a bit fearful of what the answers would be. The result is this book.

Writing this book brought me in contact with ideas, people, and places that I would have never met otherwise. My "new friends" saw me as an interesting, stimulating person. They loved my remembrances, the stories about the people my husband and I had met and our experiences on the trips to diverse countries where he had been invited to lecture; and it had always been my good fortune to be working in a company where I could take the time off to accompany him on those trips.

My new circle of friends encouraged me to do something that would never have occurred to me: to give talks about the people we met and how they colored my life. I rejected the idea at first, but they never gave up nagging me; and in the end, I finally agreed to try. When I did, the results were what they had anticipated. As for me it showed I could accomplish what I thought I never could. I was amazed at my recollections how successful my talks were and the warm response they generated.

We had met many of the famous people everyone reads about, internationally recognized Jewish leaders, like Israeli prime ministers Ben Gurion and Menachim Begin; the internationally recognized Jewish philosopher, Martin Buber; and two important Arab personalities, Muhammed Ali Jaabari, the Arab mayor of Hebron, and Saad Haadad, the Christian-Lebanese major in charge of the army in Southern Lebanon. We met many people with whom we talked and learned, first hand, their views and thoughts on a variety of subjects. I did not speak of the leaders as political figures, but as I remembered them: as warm and human personalities, well aware of their deep sense of responsibility because of how their actions and views affected their own people as well as the rest of the world. I was impressed with how strong had been their regret, since so often circumstances left them no choice but to say what they did, even when they wished they could have avoided doing so. This view is unlike the one we have of these men from newspaper articles or publications, where they are usually presented as an adjunct to the current political scene. I presented the political picture as an adjunct to these important figures and how their personalities and responsibilities influenced events.

Meeting them must have changed me, and I have often wondered how different I would have been without them. And with each of these new experiences, my view of myself expanded positively. I became more confident, expressed my feelings and ideas unhesitatingly. Could having had no college education been the reason why I always "underestimated my abilities"?

I always wonder: What brought me to where I am today? Was it the role my husband played in my life, my memories, the accident, or my refusal to "Forget it, it will go away" that made me hold on to as much of the pre-injured me as I did?

Or was it all of these? Or was it simply my good fortune that the events in my life were always so important to me that, when I needed them most, they were there to help me build a new life without my husband?

I was surprised when I realized how much I was now beginning to rely on the "new" me. Was this because the circumstances had changed, or had I changed? Was it that, with my new situation and new needs, I perceived myself differently? I had even begun to like this new me, and I wondered why. Was it because I needed help so badly that I had no choice but to be the one I could always count on? My strong need to find help did not allow me to reject it when it appeared, even when I was in doubt. I could not afford to be too proud to accept help, regardless of from whence it came. This was another decision that took me to where I am today.

In spite of the amazing progress I made, there is, however, still something missing, a part of my personality that is impossible to recoup, because it has nothing to do with memory: It has to do with feelings. Only one memory, a feeling of profound sadness, can still generate "deep" responses in me. It started about three months after the accident. The first time it happened, my husband and I were in our living room, reading, and I was relaxed and feeling good. Within a fraction of a second, like a lightning bolt, it hit me, and I went from feeling normal to being sad and depressed. It was a frightening experience, and neither of us could understand what had happened. We had no choice but to ignore it, hoping it would not happen again. But it did, over and over again, for many years. It was not until I learned I had had a brain injury that the reason for this symptom became obvious.

That living room scene has never left me. I am reminded of it each time it occurs, which is not as often, but often enough. What this injury did to me and how it affected my family will always be a part of me.

My husband lived through watching this change, and all the changes that were to follow, as he slowly watched the woman he married slip away. Neither of us ever learned what to do to reverse the process, and even today, this thought can still make me cry—not because of his death but because of what he and I missed while he was still here.

At this stage of my life, as I approach the century mark, reasons are no longer important. At times I still feel inadequate and this is to be expected. I do not, however, let it hold me back. If I did, this book, and my new life, could never have been.

My life now is relatively normal, but it is an intellectual approach to normalcy. True normal is being able to feel emotions, deeply. How one responds to circumstances is an important character trait that differentiates one person from another. I can no longer feel the normal responses that once made me *me*. They are gone, never to return. I have stopped wishing for them, and that, I suppose, is progress. Those lost responses cannot be revived. Accepting this has brought me to where I am today, living in the present, able to overcome what once were insurmountable odds, being productive, and enjoying what life has to offer. When after 30 long years the difficult search for a "new" normalcy began, success had been out of sight.

The search for the "old me" is finally over. The "new me," a combination of both old and new, seems to be in control; but as it was happening, it was so subtle I was unaware of it. How much did I have to do with this result? This answer, too, is irrelevant. Do I like the "new me?" Circumstances have given me no choice. She seems to be helping me. She has given up searching for what she cannot have, concentrating instead on showing me how to do what I need to do. Still, I have one regret: that she did not appear sooner, or, if she did, that I did not have the ability to recognize it.

Permit me a personal note: I have one more wish to fulfill—that this book will encourage others with hidden, "closed head" injuries and their families to seek help before it is too late. Symptoms become chronic all too soon, and avoiding those symptom that can be avoided is paramount to being able to return to a normal life.

EPILOGUE

How Should the Hidden Injury be Managed?

Russell C. Packard, M.D., FACP

I was asked to comment on how a concussion or closed head injury should be treated, when it happens, rather than after the symptoms become chronic. Even though I am the director of a clinic dedicated to headache and head injury, many of my new patients have for months (or in some cases years) after the injury still have persistent symptoms following mild traumatic brain injuries (or concussion). By that time they are often discouraged, depressed and desperate.

Ideally, treatment should begin in the emergency room and with the family or primary care doctor. As practice now stands, most of these patients, if they do not require admission, are sent home with advice to see their family doctor and that they will "probably be better in a few days or a week." So the seed is planted that this "mild head injury problem" shouldn't be much of a problem at all. If the patient also had a fractured arm or leg, this usually becomes the priority in the emergency department and the mild head injury may become "the hidden injury", because it isn't recognized. There is still a myth among many health care providers (and insurance companies) that a mild head injury, especially with no loss of consciousness, is not a serious problem.

Most patients, even the ones who recover more or less fully, find the recovery process to be a slow one, usually taking weeks to several months.

Treatment in the emergency room should begin with realistic goals and expectations being presented to the patient and the family. Optimism can usually be conveyed that around four out of five patients will recover, but it may take weeks or several months. The patient should also be offered a chance to see a specialist, (usually a neurologist or a neuropsychiatrist) who works specifically with head injury problems. This would allow the patient to know such a professional exists, if needed. The specialist also understands the problems the patient and family are experiencing. A referral to "a neurologist" or the family doctor may find little empathy or understanding. Many doctors do not want to deal with head injuries or post traumatic headache problems because of the associated insurance company squabbles, attorney involvement, payment problems, and decisions about work status, disability, etc.

Seeing a specialist on mild traumatic brain injury would also allow for an appropriate evaluation of the extent of the head injury. The earlier these studies are done, the more likely it is that abnormalities will be found. A slow EEG, an abnormal ENG (electronystagmogram) for the evaluation of dizziness, or abnormal evoked potentials show that the brain has either been injured or is not functioning correctly. Their neurological studies may return to "normal" long before the patient recovers. It is also important to emphasize that a normal test does not mean there has not been an injury. Too often, patients with a head injury or concussion will have had a CT head scan or MRI brain scan at the emergency room or by their family physician. They are told the scan is normal and "there is nothing wrong with you." Most of the brain injury problems from mild injuries will not show up on a CT scan or MRI because the disruption is on a cellular or physiological (brain function) level.

Even though the brain injury may not be able to be treated at the present time, treatment may be offered for headache, dizziness, muscle spasm in the neck or back and sleep problems or irritability. Medication, physical therapy and above all, an explanation and some education about this problem can go a long way toward eventual recovery and improvement. There are also some promising treatments on the horizon such as EEG biofeedback, which may improve some of the persistent cognitive dysfunction.

Involving the family from the very start can also be very helpful because they often don't understand what is wrong with the injured person. The patient (and family at times) might benefit from further counseling in regard to some of the problems that can surface after a head injury, including the physical symptoms of headache, dizziness and fatigue, cognitive difficulties with attention, concentration, thinking and memory, and emotional problems of irritability, frustration, anger and depression. We have found an outpatient support group to be very helpful to patients, who often feel terribly alone after a head injury. There may also be difficulty within the family or with the employer.

Damage to the brain damages the mind and disrupts the delicate balance we call "self." Although most patients do recover, there is the possibility that some may not. The patient has a right to be aware of the prognosis. The emphasis about outcome should be cautiously optimistic. I routinely give my patients a flyer about mild head injury and advise them <u>and</u> their family to read it. Many patients will forget details in conversations with the physician and this will be a reminder. Patients have a right to competent and compassionate treatment, even if the underlying injury cannot be cured.

Russell C. Packard, M.D. FACP

He was Professor of Neurology and Psychiatry from 2000-2005 at Texas Tech University School of Medicine and from 2005-2007 was Professor of Neurology and Psychiatry and vice-chair of psychiatry at the University of North Texas, School of Osteopathic Medicine.

Packard is so well known in his field involving headaches, head injuries and concussions, that Packard has been named one of Kipling's Who's Who in America 2010 Leading Business Professionals. He was named to Marquis Who's Who in America and named as one of Consumers Research Council of America 2009 America's Top Physicians in Neurology.

Hidden Knowledge and Power

H. Reg. McDaniel, M.D.

On a recent speaking tour I met Ethel Dimont. When she wrote asking if I would consider writing a chapter for a section entitled FROM THE TRENCHES IN THE 21st CENTURY in a manuscript concerning a closed head injury, my first thought was that I had nothing to offer and no time for this request. On lifting the cover letter, the title, in bold type, seemed to challenge me to respond to Ethel's request. Reflecting on those words, I had a wild hair that maybe I did have something to contribute to *THE HIDDEN INJURY; IT WASN'T ALL IN MY HEAD.*

Over the forty years of being a physician, I cannot count the number of times a suffering and resentful patient has exclaimed a protest to the deeper message contained in those words. A doctor, and usually it had been multiple, had failed to find a cause or a remedy for the patient's health compromise. Not being able to make a diagnosis and often unable to give relief for recurrent or persistent symptoms, the patient had been told, "It has got to be in your head," or more often, a diagnosis is made that carries that implication. These patients showed various degrees of hostility toward all doctors refusing to believe it was all in their head.

Such patients feel they have been violated and exploited in the fiscal, physical and spiritual spheres of their being. The final insult was the inference that..."I cannot help you because the problem is your fault. Through the years, I found myself developing the belief that there was a need to get past the patient's residual and earned resentment against all doctors. It is interesting to observe the response of someone who for months, years, or decades has harbored resentment against medicine and has rejected or been unresponsive to the efforts of medical professionals. In an effort to introduce a healing premise in the mind I try to tenderly massage the verbal assault on their established resentment mechanism.

Over the years I have found that the most important and powerful place for anything to be is in the mind. I know of a no more difficult or powerful seat of dysfunction to deal with. A first step is establishing communication with a health-depleted person in great need, who is resentful and distrusting of medical professionals who, for whatever reason, failed the individual.

I share a true story: "I had been following a patient for over a year after he had had cancer surgically removed from his scrotal sac. Part of the surgery required removing half of a strong visual affirmation of masculinity, his testicle. This can be as thorough an assault on one's self image for the male as having a breast removed surgically for the female. I was aware that there was a pea-sized nodule in his scrotum for some time and that it could be a benign suture granuloma, scar tissue, or a return of his cancer. It was firm, rounded, smooth-surfaced, discrete, and the size had not changed in months.

At 11:00 AM one morning, I saw him with his daughter who told me that, for months, he had continued to be active after the surgery. At 1:00 PM they had an appointment at the clinic but the surgeon who had done the surgery had retired.

A few days later, the daughter came back to see me, alone. After a long wait, she was finally able to tell me what had happened. She needed to talk to someone. At that 1:00 PM meeting, the new surgeon came in, with little ceremony and virtually making no eye contact, had her father drop his pants, with his daughter in full attendance, and showing no respect for his modesty, with a gloved hand felt his scrotum.

On finding the small nodule he exclaimed, "Oh my God, your cancer is back." He then told him they could schedule repeat surgery in the near future but they needed to do a workup to make sure it wasn't already in his lymph nodes, bones, liver or lungs.

The voice of his daughter, as she related the story, shifted from anger to crying. After that episode, with some difficulty, she had had to lead her father to the car, had to lift his legs into the car and place them in front of him. She looked back, longingly, as she related how that morning after being at my office, he had needed no assistance, carried on a normal conversation through the day until that experience in the examination room. Upon arriving home, he would not respond to her words. It was necessary to ask several neighbors to carry her father into the house. Within an hour, he was in bed and she could not get him to speak. With great emotion in her voice, she explained that he seemed in such a deep sleep that she feared it was a coma.

She could not understand and wanted to know what was happening. I told her he might have had a stroke, but I really feared he was "checking out on us." I explained that in my family there was an operative myth. It was that you get old, (my family members commonly live to their early nineties), you fall and break your hip, and you die. This has been happening for at least three generations I knew about.

In those years, no one knew about repairing a hip, or even replacing one. My two great aunts had recently followed this script and died. They believed they would die and it took three days after their fall, just like the great-grandparents, their grandparents, and their mother and father. I explained that her father does not want to struggle with another operation and the cancer treatment, knowing full well what it meant having lived through it before. He could be choosing death as an alternative.

I did suggest, however, he be examined for a stroke and if there was none, perhaps she could get a family member or a representative from a faith based church to reach him with prayer and spiritual activity, otherwise he would die. About a week later she called to inform me that no evidence of a stroke had been found by their family doctor. He died the night of the third day after he went mute.

"I relate this story to emphasize the negative or destructive power of the mind and that what we believe can and will have power over us. If such assaults, on one's health and well being can be caused by the words of a thoughtless surgeon or on beliefs in a family, can produce death, can you not see that pain, discomfort, and even disease can be brought upon our bodies by our thoughts, emotions, the careless words of others or our inability to handle negative or painful experiences in a healthy way?"

I learned of one final act of communication that transpires, and the important role it plays, when the patient interprets an act, or omission of an act, as rejection. It is not verbal. I did not recognize the power of this interaction until one day when it was brought to my attention on an especially busy day at the Fisher Institute Clinic, running back and forth between seeing patients and taking care of hospital duties.

That day I had been coming into my office through the front door and going out through the back door. Five o'clock, that evening, as I walked through the lobby to leave, six of my patients, men over 70 years of age, whom I had seen for advance cancer starting as early as one o'clock, were still sitting in that lobby. They looked forlorn. Surprised at seeing them, I asked "What's wrong? Why haven't you gone home? Is there something wrong?" Mr. Frank McMullin, looking sheepishly side to side at the others, all having dropped their eyes to the floor like little boys disappointed at being left out or forgotten but not wanting to say it hurt, spoke for the group.

His response, in a rather embarrassed manner, sheepishly said, "You were so busy you did not hug us and say goodbye." His voice quivered with emotion. I could only hope my response would help replace those hurts. I apologized, explained that it was thoughtless of me, and being called to the hospital repeatedly was not a good excuse.

As I walked forward, they lined up, one after the other. I threw my arms around their cancer tormented minds and bodies, hugged each one heartedly and this time I whispered..."I'm glad you came today, and I'm glad you waited. I'm sorry I forgot. I hope you know I love you."

After that day, I found myself wondering if what I was trying to do through the use of glyconutritional dietary supplement research was responsible for the improvements in these patients, or was it the infusion of a human being showing care for each of them and their condition. Was this act a powerful solace for isolation, disappointment, fears, or any of a number of other things that are important to each individual?

This experience, and other similar ones, taught me there is a linkage between the activity of the mind that functions in the structure of the brain, that feeds into the pineal gland, pituitary, endocrine systems, cellular and humoral defense and immune systems, as well as all other organs. Our thoughts can produce as powerful a chemical reaction in the brain as drugs do. When we remove the "human" component to medicine, we lose a powerful part of healing.

This linking action, an interaction, is expressed within and between the cells, the microscopic units of life, that make up the structure of all the organs and systems cited. The result of all the cells, trillions of cells, functioning together is what we call human, animal and complex plant life. The functioning of cells is conducted by biochemistry and it too is carried on within and between cells in an organized manner dictated by the DNA coded in our genes lined-up along chromosomes in the center of each nucleated cell.

The relationship and communication between cells is molecular, as evidenced by antibodies, cytokines (interferons, interleukins, chemokines, growth factor, etc.) and hormones. Medicine has long recognized the flow of energy along cells as found in nerve processes as evidence by the electroencephalogram (EEG) of the brains activity, cardiogram (EKG) of the heart's systemic contractions, and skeletal muscle activity as studied by electromyography (EMG).

Only in recent decades has Western Medicine begun to consider the possibility that there is a flow of energy through all cells and especially along blood vessels carrying flowing lectolytes in the blood. In India and China such knowledge is thousands of years old. Whether it is called chokra or chi, these civilizations have sought to ameliorate the flow of energy in the human body with acupuncture, manipulation, massage, pressure points, and other traditional methods.

Why the insights of Thomas Edison and Nikola Telsa have been ignored by medical science is puzzling. They used charged induction fields for supporting and restoring the flow of energy in the body that is presented in a basic physics text book as the creation of zeta potential when an electrolyte solution flows in a tube. The obvious is that blood contains a family of electrolytes that flow through a myriad of blood vessels that reach down to every cell by a network of capillaries.

Only recently has acupuncture with its ancient origin become relevant even though it has been considered as anything but fraud by Western Medicine. In my opinion, advancements will be made in energetics, the flow of charges as evidence of an energy flow, and will contribute to new advances in medicine and health. There is little doubt that what is in the mind in its relationship with the brain and nerves will control this activity. We started this response to Ethel's request with the bold printed proposed title.

THE HIDDEN POWER - is found in your mind, spirit, and cells of the human body.

Dr. Reg McDaniel

A native Texan, Dr. Reg McDaniel graduated from the University of Texas Southwestern Medical School and has spent 30 years practicing anatomical & clinical pathology, including positions as the Director of Pathology & Laboratories and Director of Medical Education at Dallas-Ft Worth Medical Center.

In 1981 he began research at Fisher Institute for Medical Research using a bean extract to stimulate the immune system. In 1985, he took the work of scientists that had isolated the active principle of the Aloe Vera plant, and conducted the first government-monitored studies in humans using this glyconutrient, aloe polymannose with unprecedented results. He devoted his attention to the potential of glyconutrients and other plant micronutrients to restore health by nutritionally supporting normal biochemistry under gene control. In 1996, the American Naturopathic Medical Association recognized his work with their "Discovery of the Year Award".

Friends or Foes It's Hard to Tell

Ethel Dimont

If Bar Associations have a role to play when clients feel they have been inadequately represented, this epilogue makes me wonder if they are living up to that role?

Though I did not expect to learn anything that would improve the situation, in an effort to try to understand what went wrong I asked to see my attorney's records. Knowing what to do when it is too late is not very helpful but those records forced me to entertain an idea that had not occurred to me until I read that file...to file a formal complaint. Less had been done than I had even surmised. No effort had been made to check the police report, the hospital records, the physicians' letters had not been questioned, checked or verified. The insurance companies ended up in the position of being able to reject any responsibility for their drivers being responsible thus keeping their settlement offer based on any data that would have held them more responsible.

I also found a $250.00 deduction on my final statement for a physician's bill I had paid. It never occurred to me I would have difficulty receiving a check from my attorney for this error. When I called this to his attention, I asked for a duplicate copy of the check so I could get a refund from the physician. Instead, I was told the case is closed...there is nothing to talk about.

Any doubt I may have had about filing a complaint with the Bar Association Grievance Committee was removed. My anger led me to include, in that complaint, not only the matter of the check but the incompetence of my attorney.

The representative of the Bar Association Grievance Committee listened to my story, read the file, and agreed that my complaint was valid on both counts. In spite of this, he managed somehow, to convince me not to file a formal complaint because "There are no other complaints against this attorney," using as a reason that he would see to it that I did get the $250.00 refund. Today, why I agreed is difficult for me to understand.

Months went by. I did not receive the check.

It was now necessary to repeat the above procedure. This time, however, I expected him to encourage me to file a formal complaint, especially since it was his letter requesting payment of the $250.00 which was ignored. Instead, the same suggestions was again made – do not file a grievance, while he again assured me I would receive the check.

When I asked the obvious question...would this episode about my complaint go into this attorney's file, the answer was a little disconcerting. "No, it would not because only formal complaints are placed in the file." I never did get an answer to the next obvious question, "What will you tell the next client who comes in with a complaint against this attorney? Would they too be told that there were no other complaints against him?" Is there any doubt about what they would say? Does this "solution" seem fair and equitable? Or shouldn't all complaints - verbal or formal - be placed in the file, perhaps with a qualifying statement? Is the Bar Association doing their job of "controlling" their members when their actions are questionable, as it was in this instance? It does not seem to me that they are.

Today it is difficult to understand why I did not file that formal complaint. I hope I did have a rational excuse for not doing so, but I wouldn't bet on it. I only remember I was too ill and too tired to fight for my rights. Perhaps the one good thing that came out of all this was that I never lost my desire to write this story so that others might avoid some of the pitfalls that they must face when professionals do not do their job.

Yes, I finally did get the $250.00.

The purpose of this story is not to condemn all attorneys. We have more than our share of good, reliable and honest ones as well as capable ones who live up to their responsibilities. I do believe, however, that there are more complaints against attorneys that are not filed than those that are. Clients are often not wise enough to know how to respond to matters they think they are not versed in nor do they have the time, money, and energy to do this without proper advice...but from whom? The grievance committee of the bar association, one would think, would be the place. But is it?

Is this case unusual? I don't think so. Details may be different but the misunderstandings and problems are the same. Many victims of a head injury cannot handle the "trying to get well" process, and the legal and financial problems that often are more difficult than mine were. My husband not only supported me with all the details, but also helped me at home. The family problems that develop when a homemaker cannot continue to do the necessary chores to keep the family functioning anywhere near the level she was able to before the injury are not easy to understand, especially when the physician insists there is no injury involved. The diagnosis, or misdiagnosis, followed us for the rest of our lives while I tried to continue to do the best I could. The price paid is incalculable.

What should I have done? From the beginning, I should have demanded written details from my attorney, questioned what was being done, and progress of the case. I should have made sure I understood clearly what his explanations meant. Nothing should have been accepted on blind faith. If I could not get this, I should have fired him.

Did the Grievance Committee do its job? I do not think so. Should they have discouraged me from filing that formal complaint? Perhaps the first time…but the second, I don't think so? Both complaints were forced on me by the attorney. Ignoring the bar association's request that he return the $250.00 which they believed he owed me should have been viewed more seriously by them than it was. Did they ever reprimand him for this "illegal" action? I doubt it.

I do not have specific advice for the legal community. Perhaps they are still working on programs to "punish," or "weed out," attorneys who are repeat offenders or just lazy. Encouraging dissatisfied clients with valid complaints who want to formalize them is one constructive first step in the right direction. When the medical profession ignores these complaints are they not responsible for encouraging malpractice suits? This has had a tremendous negative effect on that profession. Are we beginning to see signs of the same format being applied to the legal profession? It all makes one wonder.

BIBLIOGRAPHY

Alexander, D.A.
The application of the Graham-Kendall Memory-for-Designs Test to elderly normal an psychiatric groups. PMID: 5488966 [PubMed - indexed for MEDLINE] Br J Soc Clin Psychol. 1970 Feb;9(1):85-6.

Bernstein, Barbara; Kane, Robert
Physicians' Attitudes Toward Female Patients, 19. Med. Care 600 (1981) (citing G. Corea, The Hidden Malpractice (1977)., 1883 - http://tinyurl.com/2wdof2z

Berrol, Sheldon MD,
Reviewer, University of California— San Francisco

Brown, Peter Megargee
Is the Lawyer Disciplinary System Working? The Clear Tension Between Punishment Sympathy, 64 N.Y. St. B.J. 6 (1992).

Bruno, Frank J.
Professor of Social Work Research Washington University in St. Louis, St. Louis, MO

Bryant, B.; Mayou, R.; Lloyd-Bostock, S.
(1997) Compensation claims following road accidents: a six-year follow-up study. *Medicine, Science and the Law*, 37, 326-336.
(2002) Psychiatry of whiplash neck injury. *British Journal of Psychiatry*, 180, 441-448.
http://bjp.rcpsych.org/cgi/content/abstract/180/5/441

Burger, Warren E.
http://www.wmitchell.edu/lawreview/Volume30/Issue1/5Haydoc k.pdf
http://www.quotesdaddy.com/quote/248472/warren-e-burger/we-are-more-casual-about-qualifying-the-people-we

Cengage, Gale "Rorschach Technique."
Encyclopedia of Psychology. 2nd ed. Ed. Bonnie R. Strickland. 2001. eNotes.com. 2006. 19 Mar, 2010
<http://www.enotes.com/gale-psychology-encyclopedia/rorschach-technique>

Chibnall, John T., PhD
Professor, Department of Neurology & Psychiatry, St. Louis University, 1438 S. Grand Blvd., St. Louis, MO 63104
chibnajt@slu.edu
Chairman of Ford Foundation Seminar on Culture and Communication: 1953-1955. Co-Editor of Explorations magazine: 1954-1959.

Clinchot, Daniel M.
Defining sleep disturbance after brain injury - Am J Phys Med Rehabil - 01-JUL-1998; 77(4): 291-5
MEDLINE® is the source for the citation and abstract of this record

Clinchot, Dr Daniel Michael
Associate Dean, College of Medicine Clinchot, D, M; Bogner, J; Mysiw, W, J; Fugate, L; Corrigan, J. 1998. Defining sleep disturbance after brain injury.
American journal of physical medicine & rehabilitation /Association of Academic Physiatrists. Vol. 77, no. 4: 291-5.
Coghill, Robert C.
The Journal of Neurophysiology Vol. 82 No. 4 October 1999, pp. 1934-1943
Copyright ©1999 by the American Physiological Society 12345
Concussion is a trauma-induced change in mental status, with confusion and amnesia, and with or without a brief loss of consciousness.

Cooper, Astley
http://www.spartacus.schoolnet.co.uk/WcooperA.htm

Courville, C. B.
NJS - July 1954 Volume 11, Number 4
Excision of Multisaccular Supratentorial Aneurysm of Infratentorial Origin Case Report, Eldridge Campbell, M.D., Dogan Perese, M.D., and Nolton H. Bigelow, M.D., Departments of Surgery and Pathology, Albany Medical College, Albany, New York
DOI: 10.3171/jns.1954.11.4.0422

Davis, E. Marcus
Attorney at Law
http://www.emarcusdavis.com/articles/braininjuries.htm

Davis, E Marcus
"Mild to Moderate Brain Injury: A Silent Epidemic Needing To Be Heard", an article in *The Verdict* published by Journal of the Georgia Trial Lawyers Assoc. May, 1990

DeCherney, Alan H.
Obstetric & Gynecological Diagnosis & Treatment
(0639785324881): Books. 1995 Muriel Deutsch Neuropsychological
Assessment, Oxford University Press 1995

Dorland's Medical Dictionary
http://www.mercksource.com/pp/us/cns/cns_hl_dorlands_split.js
p?pg=/ppdocs/us/common/dorlands/dorland/misc/dmd-a-b-
000.htm

Emergency Neuroradiology
Springer Berlin Heidelberg, DOI: 10.1007/3-540-29941-6 ©2006
ISBN: 978-3-540-29626-3 (Print) 978-3-540-29941-7 (Online) Part
– 2 DOI: 10.1007/3-540-29941-6_11, Pg. 131-135
http://www.springerlink.com/content/h3640r/?p=16af70fd69484
64095933dde450034d8&pi=0

Erichsen, John
http://www.braceface.com/medical/Medical_Authors_Faculty/Eri
chsen_John.htm

Evans, Randolph W. MD; Tad Seifert, MD; Kailasam, Jayasree MD;
Mathew, Ninan T. MD
The Use of Questions to Determine the Presence of Photophobia
and Phonophobia During Migraine Posted: 05/21/2008;
Headache. 2008;48(3):395-397. © 2008 Blackwell Publishing

Evens R. W. MD
Headache: The Journal of Head and Face Pain
http://www3.interscience.wiley.com/journal/118518032/home?C
RETRY=1&SRETRY=0
Volume 34 Issue 5, Pages 268 – 274 **Published Online:**
18 May 2005 Copyright © 2010 American Headache Society

Falkner Wood, Sue
Just because you cannot get a diagnosis, or you are faced with a
physician who doesn't appear to understand the severity of what
you are experiencing each day, does not mean it is "in your head."
http://www.everydayhealth.com/blog/life-with-chronic-pain/is-
my-chronic-pain-all-in-my-head/

Fisher, J M; Williams, A D
Practice of forensic neuropsychology : meeting challenges in the
courtroom.
http://www.lavoisier.fr/notice/frFWOXA3AAKXWO2O.html

Frazer, J. G.
"Preface," Creation and Evolution in Primitive Cosmogonies (London, 1935), p. viii.

Gargan, William
http://news.google.com/newspapers?nid=1798&dat=19660308&i
d=C_YeAAAAIBAJ&sjid=AowEAAAAIBAJ&pg=5545,1617138

Geisel, Theodor
Of Sneetches and Whos the Good Dr. Seuss, Essays on the
Writings and life of Theodor Geisel, edited by Thomas, Fensch,
McFarland & Company, Inc., Publishers 1997 p. 79

Gide, André Paul Guillaume
(22 November 1869—19 February 1951) was a French author and
winner of the Nobel Prize in literature in 1947.
http://en.wikipedia.org/wiki/Andr%C3%A9_Gid

Greenhill, J. P. - (Jacob Pearl), 1895-1986

Green, B. L.; Saenz, D. S.
St. Louis Post-Dispatch, pp. 1&4. **(1995). Gender** differences in the
causal elements of self-efficacy on a high avoidance motor task.

Gordian Knot Theory
http://www.maa.org/devlin/devlin_9_01.html

Gronwall, D.; Wrightson, P.
The Lancet, Volume 304, Issue 7881, 14 September 1974, Pages
605-609 Gronwall, D. & Wrightson P., - Study group – J Neurol
Neurosurg Psychiatry 1981;**44**:889-895 doi:10.1136/jnnp.44.10.889
"Memory and Information Processing Capacity After Closed Head
Injury."

Gronwall, Dr. Dorothy
Delayed Recovery Of Intellectual Function After Minor Head
Injury. *The Lancet*, Volume 304, Issue 7881, Pages 605-609
http://linkinghub.elsevier.com/retrieve/pii/S0140673674919394

Gurion, David Ben
http://www.jewishvirtuallibrary.org/jsource/biography/ben_gurio
n.html

Haadad, Saad
http://middleeast.about.com/od/arabisraeliconflict/p/me081026b.
htm

Hare-Mustin, Rachel
Making a Difference: Psychology and the Construction of Gender
(Paperback - Jan. 29, 1992)
http://www.afta.org/about/rachel-t-hare-mustin-phd

Hendler, Nelson MD
Practices Psychiatry in Stevenson, Maryland. Dr. Hendler graduated with an MD 38 years ago.
Mensana Clinic, 1718 Greenspring Valley Road, Stevenson, MD 21153

Hibbard, M. R.; Bogdany, J.; Uysal, S.; Kepler, K.; Silver, J. M.; Gordon, W. A. et al.
(2000). Axis II psychopathology in individuals with traumatic brain injury. Brain Injury, 14(1), 45-61

Hinnant, Donald W. PhD
Clinical training: University of North Texas, Denton, TX and Tulane University School of Medicine

Hoffmann, Diane E. & Tarziany, Anita J.
The Girl Who Cried Pain: A Bias Against Women in the Treatment of Pain, University of Maryland School of Law. This paper is posted at http://digitalcommons.law.umaryland.edu/fac pubs/145

Hoffman, M.
The Hidden Malpractice: How American Medicine Treats Women as Patients and Professionals
27, Issue 6, Pages 1196-1198

Hyde, Thomas E.; Gengenbachm, Marianne S.-
Conservative Management of Sports Injuries, Head Trauma in Sports, Sequelae of Head Injuries pg. 339 Second Impact Syndrome (Hardcover - Apr. 20, 2007) http://tinyurl.com/38r2a9f

Jaabari, Muhammed Ali
http://www.passia.org/palestine_facts/personalities/alpha_j.htm

Jacobson, R. R.
Journal of Psychosomatic Research, The post-concussional syndrome: Physiogenesis, psychogenesis and malingering. An integrative model, Elsevier: August 1995, Copyright © 1995, Elsevier

Jennett, Bryan
Predicting Outcome In Individual Patients After Severe Head Injury
The Lancet, Volume 307, Issue 7968, 15 May 1976, Pages 1031-1034
http://tinyurl.com/2wvue9k
Jones, Michael
The Community Interpreter: A Special Case
Journal of American Medical Association. 1989:261(24):3622
Jones, Michael

The Community Interpreter: A Special Case
Published in: Australian Social Work, Volume 38, Issue 3
September 1985, pages 35 – 38
http://www.informaworld.com/smpp/title~db=all~content=t725
304176
Published in: Australian Social Work, Volume 38, Issue 3
September 1985, pages 35 – 38
http://www.informaworld.com/smpp/title~db=all~content=t725
304176

Justice Richard Sims
Third District C.A.

Kay,T.
Minor Head Injury: An Introduction for Professionals.MA:
National Head Injury Foundation. (1986)
http://www.zoominfo.com/people/Kay_Thomas_194438315.aspx

Kraus, Jess F MPH, PhD
Professor of Epidemiology and Director of the Southern California
Injury Prevention Research Center at the University of California
Los Angeles.

Kraus, Jess F.; Peek-Asa, C; Howard, J; Vargas, L; Kraus, JF
Incidence of non-fatal workplace assault injuries determined from
employer's reports in California. *J Occup Environ Med,* 39:44-50,
1997.

Leape,Lucian
Adjunct Professor of Health Policy, Department of Health Policy
and Management http://www.hsph.harvard.edu/faculty/lucian-
leape/

Leape, Lucian M.D.
JAMA, 1995, Page 1534.)
http://www.webmm.ahrq.gov/perspective.aspx?perspectiveID=28
Generally known as the father of the modern patient safety
movement in the United States.

Leriche, René
French surgeon, born October 12, 1879, Roanne, Loire; died
December 28, 1955, Cassis, near Marseille.Lezak

Maty – Boyer – Dupuytren - Cooper
http://www.neuropsychologycentral.com/.../mild_head_injury_and_pos
ttraumatic_headache.pdf%20-

Mayou, R; Bryant, B; Peveler R; Young, E.
Outcome of 'whiplash' neck injury. Injury **1996**;27(9):617-23. **1996:**

http://www.biomedexperts.com/Profile.bme/909903/Richard_
Mayou

Mayo Clinic Family Health Book
Published by William Morrow & Company
Publication date: December 1996

Mayo Clinic Family Health Book, The
Publisher: Collins; 3 Sub edition (May 6, 2003) ISBN-10:
0060002506, ISBN-13: 978-0060002503

McArthur, D.L.; Peek-Asa; C, Kraus, J.F.
Hospitalized injuries before and after the 1994 Northridge,
California earthquake. *Am J Emerg Med,* 18:361-66, 2000.

McDaniel, H. Reg. M.D.
Biographical note on Dr. Reg McDaniel, taken from the Money-
Changer interview.

McLuhan, Herbert Marshall
University of Manitoba: B. A., 1932; M. A., 1934., Cambridge
University: B. A., 1936; M. A., 1939; Ph. D., 1942. Taught at
University of Wisconsin (Madison): 1936-1937. Taught at St. Louis
University: 1937-1944. Taught at Assumption University (Windsor,
Ontario): 1944-1946. Taught at St. Michael's College, University of
Toronto: 1946-1979. Full professor: 1952.

Minnesota Multiphasic Personality Inventory (MMPI)
http://changingminds.org/explanations/preferences/mmpi.htm

Miller, Henry M.D., F.R.C.P., D.P.M.
From the Milroy Lectures for 1961, delivered before the Royal
College of Physicians of London on February 7 and 9. Published in
the British Medical Journal, April 8, 1961. p. 992-998, Journal of
Psychosomatic Research, Volume 39, Issue 6, August 1995, Pages
675-693

Miller H.
Post-traumatic headaches *Neurologic Clinics,* Volume 22, Issue 1,
Pages 237-249

Merriam-Webster
G. & C. Merriam Company, Springfield, MA, USA 1977

Munch, Shari
Gender-Biased Diagnosing of Women's Medical Complaints:
Contributions of Feminist Thought, 1970–1995, Women's Health
(Taylor & Francis – Nov. 11, 2004)
Nathan, Lauren

Obstetric & Gynecological Diagnosis & Treatment
(0639785324881): Books.

National Head Injury Foundation, USA
Minor Head Injury: An Introduction For Professionals
(permission to copy granted, 28/4/96, by author, Thomas Kay,
PhD) http://www.bianc.net/minor_head_injury.htm

Nemeth, Alexander J.
Blind Spots in the Diagnosis and Management of Minor Brain
Trauma, Medical Trial Technique Quarterly 37 (1991), pp. 478–495.

Nemeth, A. J.
(1991) Blind spots in the diagnosis and management of minor brain
trauma. Jow l of Neurology, Neurosurgery, and Psychiatry, 55, 200-
204.

Nemeth, Alexander J.
Behavior-descriptive data on cognitive, personality, and somatic
residua after relatively mild brain trauma: Studying the syndrome as
a whole, Archives of Clinical Neuropsychology, Volume 11, Issue 8,
1996, Pages 677-695, ISSN 0887-6177, DOI: 10.1016/S0887-
6177(96)80004-7.
(http://www.sciencedirect.com/science/article/B6VDJ-
4KC0W9B-4/2/930948a07f997e52729e59157c2e835a)

Nemeth, A.J.
Litigating head trauma: The "hidden" evidence of disability,
American Journal of Trial Advocacy 12 (1989), pp. 239–272.

Nemeth, A.J.
Blind spots in the diagnosis and management of minor brain
trauma, Medical Trial Technique Quarterly 37 (1991), pp. 478–495.

Nemeth, A.J.
Ambiguities caused by forensic psychology's dual identity: How to
deal with the prevailing quantitative bias and "scientistic" posture,
American Journal of Forensic Psychology 13 (1995), pp. 47–66.

Nemeth, A.J.
Why do some victims of brain trauma prematurely resume work: A
medical-diagnostic riddle, with a biopsychosocial answer (1995).

Nemeth, Alexander J.
Archives of Clinical Neuropsychology 1996 11(8):677-695;
doi:10.1093/arclin/11.8.677
New England Journal of Medicine, February 1973 *n/a*

O'Neill, Bryan
President of the Insurance Institute for Highway Safety, in 1992

Packard, Russell C. M.D., F.A.C.P.
Correspondence to Director, Neurology and Headache
Management, 5500 North Davis Highway, Pensacola, Florida
32503.
Packard, Russell C. M.D., F.A.C.P.
Adjunct Professor University of West Florida
Correspondence to Russell C. Packard, M.D., F.A.C.P.,
Director, Neurology and Headache Management, 5500 North
Davis Highway, Pensacola, Florida 32503
Page, Herbert William
Injuries of the Spine and Spinal Chord Without Apparent
Mechanical Lesion, and Nervous Shock, in Their Surgical and
Medico-Legal Aspects (Paperback - Feb. 10, 2010) pg.
Page, Herbert
Suffering and the Origins of Traumatic Memory Journal article by
Allan Young; Daedalus, Vol. 125, 1996
Parsons, LC, & Ver Beek, D.
Sleep-awake patterns following cerebral concussion. Nurs Res. 1982
Sep-Oct; 31(5):260-4.
http://www.ncbi.nlm.nih.gov/pubmed/6922465
Quote - NIH Publication #804-2478 (1984 - Page 9)
Pascal, Blaise & Fermat, Pierre de
http://www.simonsingh.net/Pierre_de_Fermat.html
Presbyterian Ministers Fund
http://wiki.answers.com/Q/Is_the_former_Presbyterian_Ministers
_Fund_now_part_of_Provident
**Philadelphia Contributionship for the Insurance of Houses from
Loss by Fire, The** http://www.contributionship.com/about.html.
Reitan Neuropsychological Laboratory
Licensed manufacturer and vendor for the Halstead-Reitan
Batteries *http://www.reitanlabs.com/catalog/default.php*
Rosman, Katherine
APRIL 13, 2010 Essay - The Power of Compassion, A doctor's
bedside manner may be the simplest innovation—and one of the
hardest to find.
Szasz, Thomas
Professor of Psychiatry Emeritus at the State University of New
York Health Science Center in Syracuse, New York, Adjunct
Scholar at the Cato Institute, Washington, D.C., author and
lecturer.

Shorter, Edward
From paralysis to fatigue: A history of psychosomatic medicine in the Modern Era. NY: The Free Press, 1992. xii + 419 pp. $24.95 (cloth) (Reviewed by Theodore M. Brown)

Silver & Kay
Neuropsychological Assessment by Muriel Deutsch Lezak 1989 (Hardcover - Mar. 2, 1995)

Sims, Richard, III
http://judgepedia.org/index.php/Rick_Sims

Strauss, Israel & Savitsky, Nathan
Am J Psychiatry 91:189-202, July 1934 doi: 10.1176/appi.ajp.91.1.189 © 1934 American Psychiatric Association
http://ajp.psychiatryonline.org/cgi/content/citation/91/1/189

Stone, Dr Jon
Consultant Neurologist and Honorary Senior Lecturer in Neurology,

Department of Clinical Neurosciences, Western General Hospital, Edinburgh EH42XU, UK; Jon.Stone@ed.ac.uk
http://www.neurosymptoms.org/cgi-bin/download.cgi

Stotland, Nada L. MD, MPH
Principles of Psychosomatic Medicine, Professor, Departments of Psychiatry and Obstetrics and Gynecology, Rush Medical College, Chicago, Illinois. Zoccolillo MS, Cloninger CR: Excess medical care of women with somatization disorder. South Med J 79: 532, 1986

Rief W,; Nanke A,; Emmerich J,; Bender A.; Zech T.
Causal illness attributions in somatoform disorders: associations with comorbidity and illness behavior. J Psychosom Res 2004;57(4):367-71
http://www.glowm.com/index.html?p=glowm.cml/section_view& articleid=409#r1

Sullivan, Katherine J. PhD, PT
Journal of Neurologic Physical Therapy: 1998 - Volume 22 - Issue 3 - pg 126-131 Functionally Distinct Learning Systems of the Brain: Implications for Brain Injury Rehabilitation

Tate, S. Shepherd
http://www.martintate.com/bio_tate.asp

Tollison, C. David PhD
Clinical Training: University of GA and Medical College of GA Journal of Occupational Rehabilitation Volume 2, Number 2 / June, 1992 Pgs. 103-107, DOT: 10.1007/BF01079017

Varney, Nils R.; Roberts, Richard J. Symonds
The evaluation and treatment of mild traumatic brain injury
Vijayan, N.; Watson,Craig
Headache: The Journal of Head and Face Pain
VL: 29 NO: 8 PG: 502-506, 1989 ON: 1526-4610 PN: 0017-8748
DOI: 10.1111/j.1526-4610.1989.hed2908502.x US:
http://dx.doi.org/10.1111/j.1526-4610.1989.hed2908502.x
Willer, Barry PhD; Leddy, John J MD; Kozlowski, Karl PhD; Donnelly,
James P PhD; Pendergast, David R EdD; Epstein, Leonard H PhD;
Clinical Journal of Sport Medicine: January 2010 - Volume 20 -
Issue 1 - pp 21-27 doi:10.1097/JSM.0b013e3181c6c22c
Wechsler Adult Intelligence Scale
http://www.mentalhelp.net/poc/view_doc.php?type=doc&id=821
9&cn=18
Weil, Andrew
Spontaneous Healing: How to Discover and Embrace Your Body's
Natural Ability, pg 78 & 709
Willer, Barry Ph.D.
Professor of Psychiatry and Rehabilitation Sciences in the UB
School of Medicine and Biomedical Sciences, is author of the Web
site and a co-author on the study.
Wittgenstein, Ludwig Josef Yohan - (1889–1951)
http://www.encyclopedia.com/doc/1G2-3404706928.html
Wolpow, Edward M.D.
Mild Head Injury: After The Fall, Harvard Health Letter, Vol. 16.
No. 6. (1991).
Wolinsky, Howard & Brune, Tom
The Serpent on the Staff: The Unhealthy Politics of the American
Medical Association. New York: Tarcher/Putnam, 1994.
http://archives.sprigg.net/booknotes/documentary/politics_wolin
sky
Zeitzer, Jamie M. PhD Friedman, Leah PhD; O'Hara, Ruth PhD
Journal of Rehabilitation Research & Development - Insomnia in
the context of traumatic brain injury Volume 46 Number 6, 2009,
Pages 827 — 836
"2-cents worth"
http://answers.yahoo.com/question/index?qid=20080730001903A
Al0NHr
Websites of interest:
http://www.medicinenet.com/head_injury/article.htm

http://www.cnn.com/2009/HEALTH/03/18/brain.injury/index.html

http://www.caregiver.org/caregiver/jsp/content_node.jsp?nodeid=396

http://www.emarcusdavis.com/articles/braininjuries2.htm

To read more on the subject of sleep visit:

http://www.nytimes.com/1997/01/05/magazine/awakening-to-sleep.html?sec=health&pagewanted=5

ABOUT THE AUTHOR

Ethel Dimont is a lecturer and author. In 1962 she co-author *Jews, God and History*, along with her husband Max I. Dimont, which received critical acclaim and has sold more than 2 million copies which the Los Angeles Times praised as, "Unquestionably the best popular history of the Jews written in the English language." Because of the success of the book, Ethel and Max, have traveled the world over, meeting with heads of state such as David Ben Gurion, Menachem Begin, Martin Buber and several other Arab Middle Eastern leaders.

In 1972 a car accident changed the focus of Ethel's life. An improperly diagnosed closed head injury caused a slow erosion of her health. She has now authored a new book, *The Hidden Injury*, which takes you through the struggles of living with a misdiagnosis, to finding the correct information that can help deal with the recovery of a head injury."

Ethel is a native of New York City, but moved to St. Louis, Missouri after her marriage to Max in 1946, (now deceased). Her daughter, son-in-law, three grandsons, two great grand-daughters and one great grandson, still live in New York City.

www.ingramcontent.com/pod-product-compliance
Lightning Source LLC
Chambersburg PA
CBHW021541200526
45163CB00014B/570